MW01243434

Energy Policy

Affirmative Action

Amateur Athletics

American Military Policy

Animal Rights

Bankruptcy Law

Blogging

Capital Punishment, Second Edition

Disaster Relief

DNA Evidence

Drugs and Sports

Educational Standards

Election Reform

Energy Policy

Environmental Regulations and Global Warming

The FCC and Regulating Indecency

Fetal Rights

Food Safety

Freedom of Speech

Gay Rights

Gun Control

Hate Crimes

Immigrants' Rights After 9/11

Immigration Policy, Second Edition

Legalized Gambling

Legalizing Marijuana

Mandatory Military Service

Media Bias

Mental Health Reform

Miranda Rights

Open Government

Physician-Assisted Suicide

Policing the Internet

Prescription and Non-prescription Drugs

Prisoners' Rights

Private Property Rights

Protecting Ideas

Religion in Public Schools

Rights of Students

The Right to Die

The Right to Privacy

Search and Seizure

Sentencing Sex Offenders

Smoking Bans, Second Edition

Stem Cell Research and Cloning

Tort Reform

Trial of Juveniles as Adults

The War on Terror, Second Edition

Welfare Reform

Women in the Military

Energy Policy

By Paul Ruschmann, J.D.

SERIES CONSULTING EDITOR
Alan Marzilli, M.A., J.D.

Energy Policy

Chelsea House
An imprint of Infobase Publishing
132 West 31st Street
New York NY 10001

Library of Congress Cataloging-in-Publication Data

Ruschmann, Paul.
Energy policy / by Paul Ruschmann.
p. cm. — (Point/counterpoint)
Includes bibliographical references and index.
ISBN 978-1-60413-333-2 (hardcover)
1. Energy policy—United States—History. 2. Energy policy—History. I. Title. II. Series.
HD9502.A4.R8 2009
333.790973—dc22 2008045706

Chelsea House books are available at special discounts when purchased in bulk quantities for businesses, associations, institutions, or sales promotions. Please call our Special Sales Department in New York at (212) 967-8800 or (800) 322-8755.

You can find Chelsea House on the World Wide Web at http://www.chelseahouse.com

Series design by Erik Lindstrom
Cover design by Keith Trego and Alicia Post

Printed in the United States of America

Bang NMSG 10 9 8 7 6 5 4 3 2 1

This book is printed on acid-free paper.

All links and Web addresses were checked and verified to be correct at the time of publication. Because of the dynamic nature of the Web, some addresses and links may have changed since publication and may no longer be valid.

POINT
COUNTERPOINT

FOREWORD

Alan Marzilli, M.A., J.D.
Birmingham, Alabama

The POINT/COUNTERPOINT series offers the reader a greater understanding of some of the most controversial issues in contemporary American society—issues such as capital punishment, immigration, gay rights, and gun control. We have looked for the most contemporary issues and have included topics—such as the controversies surrounding "blogging"—that we could not have imagined when the series began.

In each volume, the author has selected an issue of particular importance and set out some of the key arguments on both sides of the issue. Why study both sides of the debate? Maybe you have yet to make up your mind on an issue, and the arguments presented in the book will help you to form an opinion. More likely, however, you will already have an opinion on many of the issues covered by the series. There is always the chance that you will change your opinion after reading the arguments for the other side. But even if you are firmly committed to an issue—for example, school prayer or animal rights—reading both sides of the argument will help you to become a more effective advocate for your cause. By gaining an understanding of opposing arguments, you can develop answers to those arguments.

Perhaps more importantly, listening to the other side sometimes helps you see your opponent's arguments in a more human way. For example, Sister Helen Prejean, one of the nation's most visible opponents of capital punishment, has been deeply affected by her interactions with the families of murder victims. By seeing the families' grief and pain, she understands much better why people support the death penalty, and she is able to carry out her advocacy with a greater sensitivity to the needs and beliefs of death penalty supporters.

The books in the series include numerous features that help the reader to gain a greater understanding of the issues. Real-life examples illustrate the human side of the issues. Each chapter also includes excerpts from relevant laws, court cases, and other material, which provide a better foundation for understanding the arguments. The volumes contain citations to relevant sources of law and information, and an appendix guides the reader through the basics of legal research, both on the Internet and in the library. Today, through free Web sites, it is easy to access legal documents, and these books might give you ideas for your own research.

Studying the issues covered by the POINT/COUNTERPOINT series is more than an academic activity. The issues described in the book affect all of us as citizens. They are the issues that today's leaders debate and tomorrow's leaders will decide. While all of the issues covered in the POINT/COUNTERPOINT series are controversial today, and will remain so for the foreseeable future, it is entirely possible that the reader might one day play a central role in resolving the debate. Today it might seem that some debates—such as capital punishment and abortion—will never be resolved.

However, our nation's history is full of debates that seemed as though they never would be resolved, and many of the issues are now well settled—at least on the surface. In the nineteenth century, abolitionists met with widespread resistance to their efforts to end slavery. Ultimately, the controversy threatened the union, leading to the Civil War between the northern and southern states. Today, while a public debate over the merits of slavery would be unthinkable, racism persists in many aspects of society.

Similarly, today nobody questions women's right to vote. Yet at the beginning of the twentieth century, suffragists fought public battles for women's voting rights, and it was not until the passage of the Nineteenth Amendment in 1920 that the legal right of women to vote was established nationwide.

What makes an issue controversial? Often, controversies arise when most people agree that there is a problem, but people disagree about the best way to solve the problem. There is little argument that poverty is a major problem in the United States, especially in inner cities and rural areas. Yet, people disagree vehemently about the best way to address the problem. To some, the answer is social programs, such as welfare, food stamps, and public housing. However, many argue that such subsidies encourage dependence on government benefits while unfairly penalizing those who work and pay taxes, and that the real solution is to require people to support themselves.

American society is in a constant state of change, and sometimes modern practices clash with what many consider to be "traditional values," which are often rooted in conservative political views or religious beliefs. Many blame high crime rates, and problems such as poverty, illiteracy, and drug use on the breakdown of the traditional family structure of a married mother and father raising their children. Since the "sexual revolution" of the 1960s and 1970s, sparked in part by the widespread availability of the birth control pill, marriage rates have declined, and the number of children born outside of marriage has increased. The sexual revolution led to controversies over birth control, sex education, and other issues, most prominently abortion. Similarly, the gay rights movement has been challenged as a threat to traditional values. While many gay men and lesbians want to have the same right to marry and raise families as heterosexuals, many politicians and others have challenged gay marriage and adoption as a threat to American society.

Sometimes, new technology raises issues that we have never faced before, and society disagrees about the best solution. Are people free to swap music online, or does this violate the copyright laws that protect songwriters and musicians' ownership of the music that they create? Should scientists use "genetic engineering" to create new crops that are resistant to disease and pests and produce more food, or is it too risky to use a laboratory to create plants that nature never intended? Modern medicine has continued to increase the average lifespan—which is now 77 years, up from under 50 years at the beginning of the twentieth century—but many people are now choosing to die in comfort rather than living with painful ailments in their later years. For doctors, this presents an ethical dilemma: should they allow their patients to die? Should they assist patients in ending their own lives painlessly?

Perhaps the most controversial issues are those that implicate a Constitutional right. The Bill of Rights—the first 10 Amendments to the U.S. Constitution—spell out some of the most fundamental rights that distinguish our democracy from other nations with fewer freedoms. However, the sparsely worded document is open to interpretation, with each side saying that the Constitution is on their side. The Bill of Rights was meant to protect individual liberties; however, the needs of some individuals clash with society's needs. Thus, the Constitution often serves as a battleground between individuals and government officials seeking to protect society in some way. The First Amendment's guarantee of "freedom of speech" leads to some very difficult questions. Some forms of expression—such as burning an American flag—lead to public outrage, but are protected by the First Amendment. Other types of expression that most people find objectionable—such as child pornography—are not protected by the Constitution. The question is not only where to draw the line, but whether drawing lines around constitutional rights threatens our liberty.

The Bill of Rights raises many other questions about individual rights and societal "good." Is a prayer before a high school football game an "establishment of religion" prohibited by the First Amendment? Does the Second Amendment's promise of "the right to bear arms" include concealed handguns? Does stopping and frisking someone standing on a known drug corner constitute "unreasonable search and seizure" in violation of the Fourth Amendment? Although the U.S. Supreme Court has the ultimate authority in interpreting the U.S. Constitution, their answers do not always satisfy the public. When a group of nine people—sometimes by a five-to-four vote—makes a decision that affects hundreds of millions of others, public outcry can be expected. For example, the Supreme Court's 1973 ruling in *Roe v. Wade* that abortion is protected by the Constitution did little to quell the debate over abortion.

Whatever the root of the controversy, the books in the POINT/COUNTERPOINT series seek to explain to the reader both the origins of the debate, the current state of the law, and the arguments on either side of the debate. Our hope in creating this series is that the reader will be better informed about the issues facing not only our politicians, but all of our nation's citizens, and become more actively involved in resolving these debates, as voters, concerned citizens, journalists, or maybe even elected officials.

When gasoline prices exceeded $4 per gallon for the first time in 2008, sales of hybrid cars and other "green" products soared and many people questioned ongoing U.S. involvement in oil-rich Iraq. It is no surprise then that energy policies took center stage in the 2008 election. More than any other nation, the United States depends on energy for manufacturing, transportation, and household use. Once the leading producer of petroleum, the United States now depends heavily on foreign oil, and thus the U.S. role in world affairs is shaped by the nation's energy needs. This volume examines some of the science and policy behind the debate about how the U.S. government should regulate energy production and use.

Although energy conservation has been on the public agenda since the oil crises of the 1970s, which saw people desperately waiting in lines at gas stations, the debate has taken on a new intensity in the new millennium. Supporters of alternative energy, such as solar energy and wind power, are not concerned only about the environment. They warn that the world is actually running out of oil, and that unless we find ways to reduce our dependence on oil, our way of life will be threatened in the coming decades. Critics dismiss these predictions, arguing that oil supplies are plentiful, and that as new technologies develop, we will become better at finding and extracting oil. Some question the viability of alternative energy sources, saying that alternative technologies are costly, impractical, and not very versatile. Pointing out that private industries have profited by making energy-efficient products, they argue that laws requiring conservation and alternative energies are not the answer. Once alternative energy becomes efficient, they say, simple economics will encourage its use.

Energy and the Industrialized World

In recent years, energy prices have fluctuated wildly. After rising for several years, the price of crude oil spiked in 2008, reaching $147 a barrel—the highest inflation-adjusted price since the earliest days of commercial production—in July. But later that year, a worldwide economic recession triggered a collapse in oil prices. For a time in December 2008, crude was trading at less than $40 a barrel.

As prices were spiraling upward, panic set in. The investment firm Goldman Sachs warned that oil could rise to $200 per barrel within two years, and banker Matthew Simmons predicted $300-per-barrel oil within five years. High energy prices spread to other sectors of the economy, which raised concerns about further economic disruption. Automakers closed plants and laid off workers in the face of poor sales, while airlines grounded planes and imposed fuel surcharges on tickets. The price of food and a variety of other products rose, leading to concerns about inflation. The "oil shock" of 2008 also occurred in the midst of a presidential election and reopened a decades-old debate over America's energy policy.

The Energy Revolution

During the thirteenth century, residents of northeastern England discovered that certain black rocks found along the seashore, which they called "sea coles," would burn. Coal was abundant in England and would eventually play an important role in the history of energy:

> For most of history, mankind's chief energy source has been wood, and plenty of it. During the Middle Ages, Europe was densely populated with trees but not people. As a result, fuel was cheap. But thanks to a booming economy and extensive deforestation, by 1700, England was facing a major energy crisis. Or at least, much like today's America, it thought it was. Pundits rang the alarm bells about the soaring cost of wood for heating and the iron industry; the price of charcoal doubled in real terms between 1630 and 1700.[1]

There was, however, a solution to the crisis. As the price of wood rose, businessmen had an incentive to bring coal to market. Coal had a significant advantage over wood—namely, a higher "energy density," which means that a pound of coal produces more energy than a pound of wood. The use of coal meant "for the first time in human history mankind had found a huge supply of concentrated energy by means of which the energy that could be commanded by one person could be greatly increased."[2] Since then, we have used increasingly compact and energy-intense forms of power.

By the eighteenth century, England had become dependent on coal. Production, however, could not keep up with demand, in part because mines flooded with groundwater. In 1712,

Thomas Newcomen invented a steam engine that could pump water out of mines. It was expensive, noisy, and inefficient, but it enabled miners to get the coal out. Newcomen's engine changed the way we used energy. For the first time, heat energy from the burning of fuel was converted into the mechanical energy of work. Later that century, James Watt improved the design of the steam engine to the point that it could power heavy machinery. Not long afterward, James Stephenson built a locomotive to haul coal from mines to ports. Humans were now using energy to find *more* energy.

Oil and the Automobile

Human beings had known about oil for thousands of years but used only a tiny fraction of it. That began to change during the mid-nineteenth century. Oil became an alternative because there was concern that the world's coal supply could not meet demand, and because whale oil, which for centuries had been used in lighting, was becoming scarce. In 1859, a team of men led by Colonel Edwin Drake drilled the United States' first oil well. Its output was only a few barrels per day, and by today's standards was a mere curiosity. Nevertheless, it transformed America's economy. Michael Klare, a professor at Hampshire College, comments: "The United States was the first country in the world to develop a large-scale petroleum industry . . . and this industry has played a central role in sustaining the nation's economic growth for the past 145 years. Copious domestic oil output gave rise to America's first large multinational corporations, among them John D. Rockefeller's legendary Standard Oil Corporation."[3]

Early wells were shallow because oil drills could not penetrate very far below the surface. (Drake's well, for instance, was only 70 feet deep.) In 1901, however, Al and Curt Hammil, using a new rotary drill, bored more than 1,000 feet below a small hill called Spindletop in east Texas. Author Vijay Vaitheeswaran comments, "[It was] the greatest oil well that the world had ever

(continues on page 16)

Major Events in American Energy History

1859. A team led by Colonel Edwin L. Drake, an oil company executive, drills an oil well near Titusville, Pennsylvania. Drake's well ultimately leads to the development of a powerful, multinational oil industry.

c. 1880. Coal replaces wood as the nation's leading source of energy.

1901. Using a new rotary drill, Al and Curt Hammil strike oil nearly 1,000 feet below the surface at Spindletop, near Beaumont, Texas—the first major oil discovery in the United States.

1946. For the first time, the United States consumes more oil than it produces.

1951. Electric power is produced from a nuclear reactor at a government site in Idaho. Six years later, the world's first full-scale nuclear power plant, in Pennsylvania, becomes operational.

1956. M. King Hubbert, a scientist at Shell Oil Company, tells a meeting of the American Petroleum Institute that oil production follows a bell-shaped curve. Hubbert predicts that oil production in the United States will peak sometime between 1965 and 1970.

1960. Venezuela persuades Iran, Iraq, Kuwait, and Saudi Arabia to form the Organization of the Petroleum Exporting Countries (OPEC) in order to protect themselves from price swings and excessive oil company profits. In the years that follow, other major oil-producing countries in the Middle East, as well as non-Middle Eastern countries such as Indonesia and Nigeria, join OPEC.

1965. On November 9, a blackout leaves some 25 million residents of the northeastern United States and Canada without electricity for up to 12 hours. Since then, parts of the United States have experienced several major blackouts, along with numerous smaller ones.

1970. Oil production in the United States reaches a peak, confirming the Hubbert Curve.

1973. The Arab oil embargo begins when King Faisal of Saudi Arabia orders a 25 percent reduction in Saudi oil shipments to the United States. Fuel prices increase in much of the industrialized world and send America's already weak economy into a serious recession.

1975. Congress passes the first Corporate Average Fuel Economy (CAFE) standards, which require automakers to achieve an average of 18 miles per gallon (mpg) for their 1978 model-year cars, with the standard rising to 27.5 mpg by 1985. That same year, the United States creates the Strategic Petroleum Reserve, which as of October 2008 holds more than 700 million barrels—enough to meet the nation's energy needs for an estimated 33 days.

1977. In a televised address to the nation, President Jimmy Carter calls the energy crisis the "moral equivalent of war" and outlines a national energy plan that emphasizes conservation and a transition from fossil fuels to alternatives. Later that year, Congress merges a number of federal agencies into a new executive branch department called the Department of Energy.

1979. A reactor at Three Mile Island Nuclear Generating Station near Harrisburg, Pennsylvania, releases radioactive material into the environment. Although no one is killed or injured by the accident, the incident results in an almost complete halt to nuclear power plant construction in the United States.

1979. A revolution in Iran topples that nation's leader, the shah, and brings the Ayatollah Khomeini, an Islamic fundamentalist, to power. The revolution, followed by war between Iran and Iraq, creates another oil shortage that contributes to a worldwide recession.

1980. In his State of the Union address, President Carter warns that the United States will use military force if necessary to defend its interests in the Persian Gulf region. The so-called "Carter Doctrine" remains a key part of American foreign policy.

1981. President Ronald Reagan takes office and abandons Carter's energy program in favor of "deregulation." After lifting regulatory restrictions on the oil industry, the new administration relies instead on market forces to increase production and encourage conservation.

1986. The worst nuclear power plant accident in history occurs at the Chernobyl Nuclear Power Plant in the Soviet Union. The accident destroys the reactor and results in the release of a significant amount of radioactivity into the environment. Fifty-six people are killed—47 plant workers and 9 children who develop deadly thyroid cancer—and thousands of people living in the area are believed to have developed cancer as the result of radiation.

(continues)

(continued)

1988. For the first time, more than half of the oil consumed in the United States comes from foreign countries.

1990. Iraq invades and occupies neighboring Kuwait. The United States views the invasion as a threat to the region—especially to oil-rich Saudi Arabia—and assembles a coalition that ends the occupation in February 1991. The conflict, which some call a "war for oil," drives up the price of oil and helps tip America's economy into a recession.

1998. In December, the price of oil falls to $8.03 a barrel.

2001. A task force headed by Vice President Dick Cheney releases a report on national energy policy, which calls for greater production at home and making energy security a priority of U.S. trade and foreign policy.

2001. After the terrorist attacks of September 11, commentators emphasize the national-security risks stemming from U.S. dependence on oil from foreign sources—especially Saudi Arabia, the home country of 15 of the 19 men who carried out the attacks.

2005. President George W. Bush signs the Energy Policy Act of 2005, which offers tax breaks and other incentives to encourage energy production. Critics argue that the legislation is a giveaway to energy companies and does little to encourage conservation.

2007. President Bush signs the Energy Independence and Security Act. A key provision of the law raises the CAFE requirement to 27.5 mpg for 2011 model-year cars, and to at least 35 mpg by 2020.

2008. On January 2, oil trades at more than $100 a barrel for the first time. In July, the price peaks at $147 a barrel before a worldwide economic slowdown sends prices sharply downward.

(continued from page 13)

seen. Spindletop gushed for nine days and nine nights before it could be brought under control, spewing forth perhaps 100,000 barrels of oil each day."[4] The technology that recovered oil at

Spindletop made the United States the world's leading producer of oil—a position it would hold until the end of World War II. During the twentieth century, technology continued to advance. Today, it is possible to recover oil from miles below the ocean and in the remotest areas of the world.

Oil has a higher energy density than coal. It is also easier to transport, which makes it an ideal fuel for industry and especially for transportation. In fact, oil made it possible for the automobile to dominate our economy. At the time the Hammils were looking for oil at Spindletop, vehicles were replacing animals as a means of transportation. Gasoline, which is derived from oil, became the principal fuel for those vehicles because it delivered more power than rival sources such as steam and electricity. Kenneth Deffeyes, a professor emeritus at Princeton University, observes: "The automobile and the oil business were made for each other. From 1859 through 1908, the major petroleum product was kerosene for lanterns. After that, automobiles and trucks became a rapidly expanding market."[5] The United States has developed an economy that relies heavily on plentiful supplies of cheap oil. We have invested trillions of dollars in housing developments, shopping malls, office parks, and miles of roads to connect them all.

OPEC and the "Oil Shocks" of the 1970s

America's oil played a key role in the Allies' defeat of Germany, Italy, and Japan in World War II. By the end of the war, however, government officials realized that if supply-and-demand trends persisted, our country would soon become dependent on imported oil. In fact, demand for energy exceeded supply in 1946 and has exceeded supply ever since. The problem is complicated by the fact that the world's supply of oil is unevenly distributed. Most of the world's remaining supply is in the Middle East—and this has serious political and economic implications.

In September 1960, Venezuela and four countries in the Middle East met in Baghdad, Iraq, to form the Organization

The use of oil has come a long way since Edwin Drake's team drilled the first oil well in Pennsylvania in 1859. Today, oil is the lifeblood of the global economy, without which standards of living across the globe would be greatly diminished. It is also a limited energy resource.

of the Petroleum Exporting Countries (OPEC). Those countries joined forces because they believed that American- and European-owned companies were keeping the price of oil too low and making excessive profits at their expense. OPEC began as an informal bargaining unit, but in the early 1970s it began using its oil wealth to achieve political goals. On October 17, 1973, OPEC's Arab members imposed an oil embargo on the United States and other Western countries to punish them for supporting Israel in the Yom Kippur War against Egypt and Syria.

The boycott led to the 1973–1974 oil crisis, the first fuel shortage the United States experienced since World War II. The price of a gallon of gas quadrupled, and shortages persisted even after the embargo was formally lifted in March 1974. Drivers faced long lines to fill up, and many gas stations either ran out or limited how much a customer could buy. The U.S. government reacted quickly to the crisis. President Richard Nixon signed emergency legislation that authorized the government to regulate the price, allocation, and marketing of oil. He also named an "energy czar" to oversee the nation's energy policy. Congress passed a number of conservation measures, including a national 55 mph speed limit and fuel-economy standards for new cars. The federal government even printed gasoline-ration coupons to be used if the crisis got worse.

Energy remained a serious concern for the rest of the decade. High energy prices further weakened an already struggling economy. The result was "stagflation"—sluggish economic growth at the same time that the price of goods and services rose sharply. On April 18, 1977, President Jimmy Carter called the nation's energy crisis "the moral equivalent of war" and told Americans:

> I know that some of you may doubt that we face real energy shortages. The 1973 gasoline lines are gone, and our homes are warm again. But our energy problem is worse tonight

> than it was in 1973 or a few weeks ago in the dead of winter. It is worse because more waste has occurred, and more time has passed by without our planning for the future. And it will get worse every day until we act.[6]

The president outlined a plan to lessen the country's dependence on foreign oil. His plan relied heavily on conservation and on developing substitute sources of energy.

Two years later, the nation faced the 1979 oil crisis. It began when a revolution forced Iran's leader, the shah, to flee his country, and brought an Islamic fundamentalist, Ayatollah Khomeini, to power. The revolution, followed by a war between Iran and Iraq, hampered oil production in the region. Even though other OPEC countries increased their output to make up for the shortfall, worldwide production nevertheless fell and the economies of oil-consuming countries went into recession. Americans once again experienced fuel shortages and long lines to fill up. The deteriorating political situation in the Middle East—which included Americans being held hostage by Iranian revolutionaries—led President Carter to tell the nation: "Let our position be absolutely clear: An attempt by any outside force to gain control of the Persian Gulf region will be regarded as an assault on the vital interests of the United States of America, and such an assault will be repelled by any means necessary, including military force."[7] Presidents since then have made it clear that they would send American forces to the Middle East to protect our access to oil.

The Rise and Fall of Oil Prices

Energy prices stabilized during the 1980s. A slowdown in economic activity, along with conservation measures in oil-consuming countries, reduced demand for oil. At the same time, the high price of oil brought about a switch to alternatives such as nuclear power and natural gas. Oil from new sources, such as the North Sea off the British coast and the North Slope of Alaska, arrived on the world market. In addition, President Ronald Reagan lifted price controls on oil, which many blamed for aggravating the shortages of the 1970s. Increased supply and lower demand weakened OPEC, whose members began quarreling. In 1986, the world price of oil collapsed, falling from $27 a barrel to less than $10.

With one exception, the rest of the twentieth century was marked by relatively stable oil prices. That exception was the Persian Gulf War of 1990–1991, which resulted from Iraq's invasion of Kuwait. The United States responded by assembling a coalition of countries—which included Saudi Arabia and several other Arab countries—that drove the Iraqis out of Kuwait. Some call the Persian Gulf War the world's first war fought purely for oil and warn of more such conflicts to come. The price of oil stabilized after the war, then dropped dramatically in 1998 because of a number of factors, including unseasonably warm weather and economic turmoil in Russia and southeast Asia. By December of that year, the price of oil fell to about $8 a barrel, and it appeared that there would be no end to cheap energy.

Energy Crises Return

By the turn of the century, energy had once again become a concern. California experienced shortages of electric power that forced power companies to implement "rolling" blackouts. President George W. Bush, who took office in 2001, warned Americans that national security depended on the availability of energy. He assembled a National Energy Policy Development Group, headed by Vice President Dick Cheney. The panel's report compared the energy shortage with those of the 1970s: "The effects are already being felt nationwide. Many families face energy bills two to three times higher than they were a year ago. Millions of Americans find themselves dealing with rolling blackouts or brownouts; some employers must lay off workers

(continues on page 24)

M. King Hubbert's Remarks on Peak Oil

In March 1956, M. King Hubbert, a Shell Oil Company geologist, spoke to a regional meeting of the American Petroleum Institute in San Antonio, Texas. At the time his remarks* drew little attention from the mainstream media. Many within the industry disagreed with him. Today, however, most experts believe that his analysis was sound and that we are nearing a peak in oil production—if we have not already passed it.

Hubbert reminded his audience that it took 500 million years to turn plants and animals into fossil fuels and that the supply of those fuels cannot be replenished once it runs out. As to *when* fossil fuels would run out, Hubbert relied on what he called the best record of their extraction history—namely, production statistics. He presented charts that depicted production of crude oil, natural gas, and coal, and observed: "Each curve starts slowly and then rises more steeply until finally an inflection point is reached after which it becomes concave downward." Hubbert pointed out that, during the initial stages, the rate of production of fossil fuels tends to increase exponentially with time. He also noted, however, that "although production rates tend initially to increase exponentially, physical limits prevent their continuing to do so."

Hubbert then presented charts showing that the production of a finite resource follows a bell-shaped curve. He remarked, "If we knew the quantity initially present, we could draw a family of possible production curves, all of which would exhibit the common property of beginning and ending at zero, and encompassing an area equal to or less than the initial quantity." Production curves from states that once produced large amounts of oil, such as Ohio and Illinois, in fact followed Hubbert's bell-shaped curve.

Hubbert estimated the world's crude oil reserves at 1.25 trillion barrels and America's reserves at 150 billion barrels. He went on to say:

> Since the cumulative production is already a little more than 50 billion barrels, then only [100 billion barrels] are available for future production. Also, since the production rate is still increasing, the ultimate production peak must be greater than the present rate of production and must occur sometime in the future. At the same time it is impossible to delay the peak for more than a few years and still allow time for the unavoidable prolonged period of decline due to the slowing rates of extraction from depleting reservoirs.

> With due regard for these considerations, it is almost impossible to draw the production curve based on an assumed production of 150 billion barrels in any manner differing significantly from [a bell-shaped curve], according to which the curve must culminate at about 1965 and then must decline at a rate comparable to its earlier rate of growth.

He added that, even if America's total reserves were 200 billion barrels, "the date of culmination is retarded only until about 1970."

For the world as a whole, Hubbert predicted:

> On the basis of the present estimates of the ultimate reserves of petroleum and natural gas, it appears that the culmination of world production of these products should occur within about half a century.... This does not necessarily imply that the United States or other parts of the industrial world will soon become destitute of liquid and gaseous fuels, because these can be produced from other fossil fuels which occur in much greater abundance. But it does pose a national problem of primary importance, the necessity, both with regard to requirements for domestic purposes and those for national defense, of gradually having to compensate for an increasing disparity between the nation's demands for these fuels and its ability to produce them from naturally occurring accumulations of petroleum and natural gas.

Hubbert saw nuclear power as a possible long-term solution to oil and natural gas depletion. He estimated that the world's reserves of uranium—the element used to produce the nuclear reaction that generates electricity—could provide 10 times as much power as fossil fuels. With that in mind, Hubbert closed on a somewhat hopeful note:

> There is promise, however, provided mankind can solve its international problems and not destroy itself with nuclear weapons, and provided the world population (which is now expanding at such a rate as to double in less than a century) can somehow be brought under control, that we may at last have found an energy supply adequate for our needs for at least the next few centuries of the "foreseeable future."

* M. King Hubbert, "Nuclear Energy and the Fossil Fuels." Presented before the Spring Meeting of the Southern District Division of Production, American Petroleum Institute, March 7–9, 1956. Houston: Shell Development Company, 1956.

(continued from page 21)
or curtail production to absorb the rising cost of energy. Drivers across America are paying higher and higher gasoline prices."[8] It recommended that the federal government take steps to encourage greater production of energy.

The issue of global warming has further intensified the debate about energy policy. Scientists have concluded that carbon dioxide, which is a byproduct of burning fossil fuels, traps solar energy that otherwise would radiate back into space, and that these so-called "greenhouse gases" (GHGs) are a leading contributor to carbon dioxide emissions. In 2007, the Intergovernmental Panel on Climate Change (IPCC), a scientific organization established by the United Nations, warned: "Continued GHG emissions at or above current rates would cause further warming and induce many changes in the global climate system during the 21st century that would very likely be larger than those observed during the 20th century."[9] To combat global warming, the world community drew up the Kyoto Protocol, a treaty that commits the world's industrialized countries to curb their GHG emissions. The United States never agreed to Kyoto because of concerns that it would force the nation to adopt a conservation program that would hobble its economy.

The Debate About Energy Policy

On March 8, 1956, a Shell Oil Company geologist named M. King Hubbert told a group of oil company employees that the production of oil, as well as natural gas and coal, followed a predictable pattern: "Each curve starts slowly and then rises more steeply until finally an inflection point is reached after which it becomes concave downward."[10] In the bell-shaped "Hubbert curve," peak production occurs when half the supply has been recovered. After that, production falls off, gradually at first, then at an increasing rate. Hubbert predicted that oil production in the continental United States would peak between 1965 and 1970, and would peak worldwide during the early twenty-first century.

Although Hubbert's theory was not well received at first, most analysts now agree with it. However, they disagree as to *when* peak production will actually occur. Because oil and natural gas will eventually run out, energy policy raises other contentious issues. Today, there is considerable debate about what substitutes, if any, can be found for oil and gas. Many call for greater use of renewable sources such as solar or wind power. They point out that such renewable energy sources will not run out, are not controlled by foreign governments, and emit fewer greenhouse gases. Others, however, contend that renewables are no cure-all: They are more expensive and less reliable than conventional fuels, and there are serious problems associated with recovering and transporting the energy that they produce.

There is also disagreement as to what role government should play. Advocates of regulation contend that market forces tend to produce "quick fixes" at a time when a long-term approach is needed, and insist that only the government can implement a comprehensive program to move the nation from fossil fuels to renewables. On the other hand, supporters argue that market forces give entrepreneurs the incentive to find new sources of energy and develop more efficient ways of using the energy that is already available. They add that, in the past, humans found alternatives when they were forced to.

Summary

Until a few centuries ago, human beings relied on renewable energy sources. That all changed with the Industrial Revolution, which resulted in our use of fossil fuels: coal, then oil and natural gas. Demand for those fuels has grown dramatically. For much of the twentieth century, the United States was the dominant producer of oil, but other countries have since surpassed us. Today, the largest reserves are in the Middle East, which is the source of ongoing conflict. Instability in that region led to two major energy crises during the 1970s, which forced our government to look for ways to reduce consumption and find substitute fuels. After a period of stability, energy prices rose sharply in recent years before falling once again. The "oil shock" of 2008, caused worldwide political and economic instability, prompting lawmakers to make energy a top national priority. There is considerable debate as to how much oil is left, whether renewables are a workable substitute, and what role the government should play in shaping our energy policy.

Energy Depletion Is a Serious Threat

In his book *The End of Oil*, Paul Roberts describes what happened when the oil ran out at Sand Island in the central Asian republic of Azerbaijan:

> Twenty years ago, this three-hundred-acre island was the toast of the Soviet oil industry, with row after row of gushing wells and thick pipelines crossing the water to refineries in Baku. Then oil production hit its natural peak, the flow subsided, and Sand Island fell into the kind of profound industrial decay that Hollywood spends millions trying to replicate. Rusting pipelines line the roads. Empty buildings, some still sporting the red Soviet star, lean at odd angles. Old barrels, bits of broken machinery, and permanently parked trucks litter the grounds, while just offshore a line of gigantic rust-colored oil

> derricks, most of them abandoned, marches away toward the horizon.[1]

Some believe that the oil depletion that left Sand Island a wasteland is about to happen elsewhere in the world. The result, they warn, will be economic and political disruption.

The peak oil theory has been confirmed.

Although few people believed him at the time, M. King Hubbert's "peak oil" theory has been confirmed. Today, most oil analysts believe that production follows the bell-shaped Hubbert curve. The Oil Depletion Analysis Centre explains why that happens: "Oil production in a given country tends to go into decline at about the half-way point because of a combination of falling pressure in the underground reservoirs, and the fact that oil companies usually discover and exploit the largest oil fields first."[2] As supply dwindles, recovering the remaining oil becomes increasingly expensive. Eventually, the amount of energy needed to recover the oil exceeds the energy in the oil itself. At that point, it no longer makes sense to recover the oil, no matter what the price.

Oil production of the United States peaked in 1970, as Hubbert predicted. In his 1956 remarks, Hubbert made another prediction: "If the world should continue to be dependent upon the fossil fuels as its principal source of industrial energy, then . . . [o]n the basis of the present estimates of the ultimate reserves of petroleum and natural gas, it appears that the culmination of world production of these products should occur within about half a century."[3] In other words, worldwide production is now close to peaking.

Some experts, such as oil billionaire T. Boone Pickens, believe that the worldwide peak has arrived, and others believe that it is close at hand. In 2004, Michael Klare, a professor at Hampshire College, noticed two early-warning signs of a peak: Royal Dutch

Shell lowered its estimate of proven reserves by 20 percent, and oil industry experts concluded that Saudi Arabia was exhausting its reserves faster than previously assumed. The authors of the Hirsch Report, a study of oil depletion funded by the U.S. Department of Energy, explain what has been happening: There are no more "Super Giants," reservoirs where oil is the easiest to find, the most economical to develop, and the longest lived. The last Super Giants were found in 1967 and 1968. All the oil that has been found since then comes from smaller reserves, whose oil is more difficult to recover.

There are signs that we are approaching "peak gas" as well. As recently as 2001, a number of credible organizations were optimistic about the ready availability of natural gas. Now, experts warn that North American gas production is headed downward, likely for good. The recent increase in the price of gas reflects concerns about supply. After years of stability, the wellhead price (the value of oil at the mouth of the well) rose from $2.86 per thousand cubic feet in April 2000 to $8.94 in April 2008. The apparent peaking of gas production led the authors of the Hirsch Report to argue, "If experts were so wrong on their assessment of North American natural gas, are we really comfortable risking that the optimists are correct on world conventional oil production, which involves similar geological and technological issues?"[4]

Energy depletion is unavoidable.

Supply is half of the energy-depletion equation. The other is demand, which has been increasing. In 2001, the National Energy Policy Development Group estimated that worldwide oil consumption would grow by 2.1 percent per year between 2001 and 2020. Significantly, the panel estimated that demand would grow three times as fast in developing countries as in the industrialized world. India and especially China account for much of the increased demand, but other countries are likely to follow. Paul Roberts explains:

Nearly three billion people in effect live outside the modern energy system: they subsist on wood and other biomass and have little influence on the global dynamics of supply and

Conclusions of the Hirsch Report

The U.S. Department of Energy awarded Science Applications International Corporation a contract to study the possibility that oil production will peak and what should be done to prepare for it. The result of that study is the so-called "Hirsch Report,"* named for lead author Robert L. Hirsch. The report's major conclusions were:

1. We do not know when peak oil will occur.
2. In contrast with past oil crises, the problems associated with peak oil will be long-lasting.
3. Peak oil will create a severe liquid fuels problem for the transportation sector.
4. Peak oil will result in dramatically higher oil prices, which will cause protracted economic hardship in the United States and the world.
5. In developed nations, problems related to peak oil will be serious. In developing nations, peaking problems have the potential to be much worse.
6. Mitigating the effects of peak oil will require at least 10 years of intense, expensive effort.
7. More efficient energy use, by itself, will not solve the problems caused by peak oil. It will also be necessary to produce large amounts of substitute liquid fuels.
8. Governments will be required to take action because the economic and social implications of peak oil would otherwise be chaotic.

Hirsch and his coauthors expressed the opinion that oil production was more likely to peak in 10 years than in 30 years—a serious problem given their conclusion that it will require at least 10 years to prepare for the peak. They added, however, that a number of "wild cards" would determine when the actual peak would arrive. Factors that might delay peaking include these:

- Middle East oil reserves are much higher than publicly stated.
- A number of new Super Giant oilfields are found and brought into production well before peak oil might otherwise have occurred.

demand. To begin to bring even a fraction of those people up to modern energy standards—by providing them with coal-fired power plants, for example, or steady supplies of diesel or

- High world oil prices for a sustained period lead to a higher level of conservation and energy efficiency, and encourage the oil industry to construct substitute fuel plants well before production peaks.
- The United States and other consuming nations impose more stringent fuel efficiency standards well before peak oil arrives.
- World economic and population growth slows, and future demand for oil is much less than anticipated.
- China and India impose strict fuel-economy standards for vehicles and energy-efficiency requirements for machinery and appliances, thus reducing the rate at which their consumption grows.
- Huge new reserves of natural gas are discovered, a portion of which is converted to liquid fuels.
- Some kind of scientific breakthrough comes into commercial use, reducing oil demand well before production peaks.

On the other hand, the following factors might worsen the problem of oil peaking:

- Middle East reserves are much less than stated.
- Terrorism either stays at current levels or increases, and terrorists concentrate on damaging oil production, transportation, refining, and distribution.
- Political instability in major oil-producing countries results in unexpected, sustained worldwide oil shortages—or, even worse, large-scale, sustained Middle East political instability hinders oil production.
- The price of oil hides the fact that production has peaked, and, as a result, governments delay efforts to reduce consumption and find alternatives.
- Consumers demand even larger, less fuel-efficient vehicles.
- Environmental challenges hinder the expansion of energy production, resulting in shortages of energy sources other than liquid fuels.

*Robert L. Hirsch, Roger Bezdek, and Robert Wendling, *Peaking of World Oil Production: Impacts, Mitigation and Risk Management*. San Diego: Science Applications International Corporation, 2005.

stove fuel—would add enormous stress to the global energy system. To bring all of them along would change the world in ways we have trouble imagining.[5]

The 1980s and 1990s have been called "a golden age of oil-field technology development" in which energy companies used advances such as three-dimensional seismic analysis to find oil and drilling equipment capable of reaching oil miles below the surface. No amount of technology, however, can increase the remaining supply. When oil companies began drilling, they first concentrated on the easiest-to-reach, cheapest oil. That oil was underneath dry land, close to the surface, and, because it was under pressure, easily extracted. That oil was also "light," which meant it flowed easily, and "sweet," which meant it had low sulfur content and thus was easy to refine. As that supply ran out, oil companies were forced to look beneath the oceans, in remote parts of the Earth, and in smaller fields with lower-quality oil.

According to the *BP Statistical Review of World Energy*, worldwide oil production has increased only modestly in recent years, from about 60 billion barrels in 1981 to about 80 billion barrels in 2006. Even today's high prices are unlikely to bring about a large increase in production. The authors of the Hirsch Report ask: "If higher prices did not bring forth vast new supplies of North American natural gas, are we really comfortable that higher oil prices will bring forth huge new oil reserves and production, when similar geology and technologies are involved?"[6]

In this country, the problem of depletion is even more acute because we began extracting oil much earlier than most other countries. Roscoe Bartlett, a U.S. representative from Maryland, recently said on the floor of the House of Representatives: "The U.S. has drilled more oil wells than all the rest of the world combined. There are about four times as many oil wells in the Gulf of Mexico, about 4,000 in all, as there are in all of Saudi Arabia. In spite of finding oil in Alaska and in spite of finding oil in the Gulf of Mexico, we are now producing about half the oil we did in 1970."[7] Bartlett added that, even though the United States has only 2 percent of the world's oil reserves, we are pumping 8 percent of the world's oil. As a result, our supply will run out faster than the rest of the world's. That is true even if Congress opens areas such as the Arctic National Wildlife Refuge (ANWR) to drilling, in addition to the outer continental shelf. Professor Deffeyes observes: "Of course, every little bit helps, but ANWR is a very little bit. World oil consumption is roughly 25 billion barrels per year; 5 billion barrels from ANWR would postpone the world decline for two or three months."[8]

Energy shortages will devastate our economy.

America consumes 25 percent of the world's crude oil and 43 percent of its gasoline. Oil accounts for the vast majority, 95 percent or more, of the energy we use in transportation. We also use oil to make a variety of products including plastics, fertilizer, pharmaceuticals, and electronic components. Colin Campbell, who spent many years in the oil industry, observed: "The fundamental driver of the 20th century's economic prosperity has been an abundant supply of cheap oil."[9] Now that oil is no longer guaranteed to be cheap, our prosperity is in danger.

During the second half of the twentieth century, the United States increasingly relied on imported oil. As of 2008, two-thirds of the U.S. oil supply comes from other countries. In addition to posing national-security risks, which will be discussed later in this book, dependence on foreign oil is a drag on our economy. In 2002, the Union of Concerned Scientists (UCS) noted that foreign oil cost the American economy an estimated $7 trillion during the previous three decades—as much as this country paid

(continues on page 36)

Energy Policy Words and Phrases

Alternative fuels. Materials or substances, other than conventional fuels, that are used as fuels. They include biodiesel, chemically stored electricity (batteries and fuel cells), hydrogen, and biomass sources.

Barrel. The unit of measurement used in the oil industry. One barrel of oil is equal to 42 gallons.

Biomass. Organic, nonfossil material of biological origin that is a renewable source of energy.

British thermal unit (BTU). The amount of heat needed to raise the temperature of one pound of water by one degree Fahrenheit.

Coal. A combustible rock that consists mostly of carbon-rich material. Coal is formed from plant remains that have been compacted, hardened, and chemically altered as the result of heat and pressure for millions of years.

Cogeneration. The production of electricity along with another form of useful energy such as heat or steam.

Conventional fuels. Fossil fuels and nuclear materials.

Corporate Average Fuel Economy (CAFE) standards. A federal energy-conservation measure requiring automakers to meet fuel-economy standards for passenger cars. Those standards are expressed in miles per gallon (mpg).

Crude oil. A mixture of hydrocarbons that exists in liquid phase in natural underground reservoirs and remains liquid after it is pumped out.

Electricity generation. The process of transforming other forms of energy into electric energy. The amount of electric energy produced is commonly expressed in kilowatt-hours (kWh), or 1,000 watts of power expended for one hour.

Energy. The capacity for doing work. There are several forms of energy, some of which are easily convertible into another form that can do work. Most of the world's convertible energy comes from fossil fuels that are burned to produce heat, which is then used as a transfer medium to mechanical or other means in order to accomplish tasks. Electric energy is usually measured in kilowatt-hours, whereas heat energy is usually measured in British thermal units.

Energy efficiency. The ability to use less energy to produce the same amount of useful work or services.

Energy intensity. The amount of energy used to produce one dollar's worth of goods or services. It is possible for energy intensity to decrease while overall energy consumption rises.

Energy source. A substance that supplies heat or power.

Fossil fuel. A fuel formed in the Earth's crust, such as oil, coal, or natural gas. Most energy used in the world today comes from fossil fuels.

Fuel. A substance used to store energy in a form that is stable and easily transportable to the user. When the user consumes the fuel, energy is released, usually in the form of heat.

Geothermal energy. Energy from the internal heat of the Earth. The heat is found in rocks and fluids at various depths and can be extracted by drilling or pumping.

Hydrocarbon. An organic chemical compound of hydrogen and carbon. Hydrocarbons can be solid, liquid, or gas.

Hydropower. The production of electricity from the energy of falling water.

Internal-combustion engine. An engine that provides power as the result of the combustion of fuel and air in a confined space. The combustion produces expanding hot gases that cause the engine's solid parts to move. Most vehicles are powered by internal-combustion engines that run on gasoline or some other by-product of oil.

Methane. A hydrocarbon gas that is the major component of natural gas.

Natural gas. A gaseous mixture of hydrocarbon compounds, primarily methane, delivered via pipeline for consumption. It is used as a fuel to generate electricity, run machinery, and heat homes and businesses.

Nuclear power. Electricity generated by an electric power plant whose turbines are driven by steam generated in a reactor by heat from the splitting of the nuclei of atoms.

Organization of the Petroleum Exporting Countries (OPEC). An organization formed in 1960 to coordinate the policies of oil-producing countries, stabilize prices and supply, and provide a steady income to oil producers. Current OPEC members are Algeria, Angola, Ecuador, Iran, Iraq, Kuwait, Libya, Nigeria, Qatar, Saudi Arabia, the United Arab Emirates, and Venezuela.

(continues)

(continued)

Peak oil. The point at which the maximum rate of worldwide oil extraction occurs. After the peak, the rate of production goes into an irreversible decline. This theory, advanced by M. King Hubbert in 1956, has been used to predict the peaking and decline of oil production in the United States and other countries.

Proven reserves. The estimated amount of oil or natural gas that, according to geological and engineering data, is likely to be recovered from known reservoirs.

Refinery. An installation that manufactures finished fuels from oil, liquid natural gas, or other hydrocarbons.

Renewable energy. Energy obtained from sources that are essentially inexhaustible. (Fossil fuels are not renewable because there is a finite supply of them.) Renewable sources of energy include hydroelectric power, wood, waste, geothermal, wind, and solar energy.

Reserves. The estimated amount of oil or natural gas in the ground.

Spot price. The price for a one-time purchase of a product on the open market for immediate delivery at the point of purchase.

Transportation sector. Machines that move people and products. The sector includes motor vehicles, railroads, aircraft, and ships. It also includes natural gas pipelines.

Wind energy. The energy of wind converted into mechanical energy by wind turbines (blades rotating from a hub) that drive generators to produce electricity.

Wood energy. Energy obtained from wood products that are used as fuel. Those products include wood chips, bark, sawdust, and charcoal.

Work. The transfer of energy from one object to another, causing the second object to move. The amount of work done depends on how far the second object moves.

(continued from page 33)

on the national debt during that period. Since then, the total has grown larger still. Furthermore, there is no assurance that other countries will continue to sell us as much oil as we demand. Some countries, such as Iran, have used oil as an economic weapon and might do so again. Others, such as Mexico, might stop exporting because they need the oil to meet growing demand at home.

A small drop in oil production can trigger a huge price increase. The UCS explains:

> The political instability of the Persian Gulf has caused three major price shocks over the past 30 years. The Iraqi invasion of Kuwait in 1990 took an estimated 4.6 million barrels per day out of the global oil supply for three months. The Iranian revolution reduced global oil supplies by 3.5 million barrels per day for six months in 1979, and the Arab oil embargo eliminated 2.6 million barrels per day for six months in 1973. . . . In each of these cases, the world oil supply dropped only about 5 percent . . . but world oil prices doubled or tripled.[10]

Making matters worse is that there is little, if any, surplus production capacity in the world. Paul Roberts observes: "[I]t's now clear that even the Saudis lack the physical capacity to bring enough oil to desperate consumers. As a result, oil markets are now so tight that even a minor disturbance—accelerated fighting in Iraq, another bomb in Riyadh, more unrest in Venezuela or Nigeria—could send prices soaring and crash the global economy into a recession."[11]

Even if the world's oil-producing regions remain peaceful, the arrival of peak oil will inevitably—and perhaps in the span of a few years—cut production by a greater amount than the events that triggered the oil shocks of the 1970s. Roberts believes that the consequences will be devastating:

> An inflationary ripple effect would set in. As energy became more expensive, so would such energy-dependent activities as manufacturing and transportation. Commercial activity would slow, and segments of the global economy especially

> dependent on rapid growth—which is to say, pretty much everything these days—would tip into recession. The cost of goods and services would rise, ultimately depressing economic demand and throwing the entire economy into an enduring depression that would make 1929 look like a dress rehearsal and could touch off a desperate and probably violent conflict for whatever oil supplies remained.[12]

Author James Kunstler is even more apocalyptic. He warns: "Peak is quite literally a tipping point. Beyond peak, things unravel and the center does not hold. Beyond peak oil, all bets are off about civilization's future."[13] Kunstler foresees a "long emergency" in which globalization comes to a halt, standards of living fall dramatically, and social order breaks down.

We cannot afford to wait any longer.

The authors of the Hirsch Report emphasized the seriousness of peak oil: "The world has never faced a problem like this. . . . Previous energy transitions (wood to coal and coal to oil) were gradual and evolutionary; oil peaking will be abrupt and revolutionary."[14] American policymakers have known for decades that oil is a finite resource. Nevertheless, they have done little to lessen our dependence on fossil fuels, especially imported oil. Deffeyes comments: "After you drive a car off a cliff, it's too late to hit the brakes. In effect, we have gone over the edge of the cliff. In addition to M. King Hubbert's generalized warning, over the last twenty years a dozen different authors predicted that world oil production would peak and start a permanent decline during the 2000–10 decade."[15]

Even if pessimists such as Pickens are wrong about its timing, we know that oil production *will* peak. Making matters worse is that we probably will not know for certain that the peak has occurred until after the fact, when it will be too late to act. As the authors of the Hirsch Report explain:

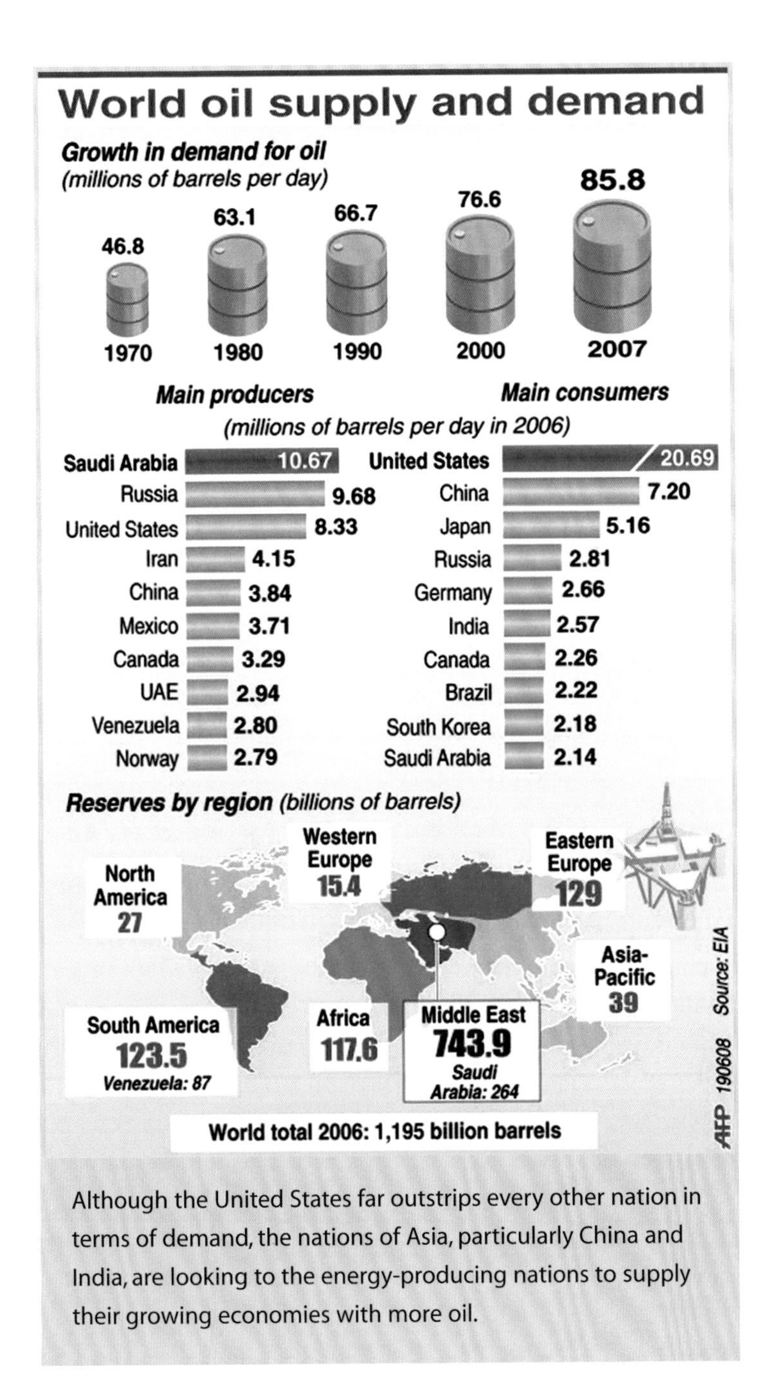

Although the United States far outstrips every other nation in terms of demand, the nations of Asia, particularly China and India, are looking to the energy-producing nations to supply their growing economies with more oil.

- Waiting until world oil production peaks before taking crash program action leaves the world with a significant liquid fuels deficit for more than two decades.
- Initiating a mitigation crash program 10 years before world oil peaking helps considerably but still leaves a liquid fuels shortfall roughly a decade after the time that oil would have peaked.
- Initiating a mitigation crash program 20 years before peaking appears to offer the possibility of avoiding a world liquid fuels shortfall for the forecast period.[16]

Some, such as Roberts, consider peak oil a more pressing concern than even global warming, another grave threat to humans that has received considerable media attention. Roberts observes: "Comforting as it might be to imagine the decline of our energy economy as a long-term process—with oil supplies peaking in 2025, say, or sea levels rising by 2050—there are fewer and fewer reasons to believe that our overtaxed energy system won't have begun to collapse long before then."[17]

Finally, the oil shocks of the twentieth century did little to prepare us for what is to come. The authors of the Hirsch Report remind us that "over the past 30 years, most economic studies of the impact of oil supply disruptions assumed that the interruptions were temporary and that each situation would shortly return to 'normal,' " and added that, since 1970, "most large oil price increases were eventually followed by oil price declines, and, since these cycles were expected to be repeated, it was generally felt that 'the problem will take care of itself as long at the government does nothing and does not interfere.' "[18] On the other hand, when production peaks, the result will be a *permanent* disruption of supply that will grow worse over time.

Summary

M. King Hubbert's prediction that oil production follows a bell-shaped curve has been confirmed. Production in America peaked in 1970, and neither technology nor high prices will restore production to its peak level. There are signs that worldwide production of oil as well as natural gas are approaching a peak, if a peak has not already occurred. Meanwhile, demand for energy continues to grow, especially in the developing world. The likely result will be a worldwide shortage that devastates countries whose economy depends on cheap energy. Unlike previous oil shocks, the crisis that follows peak oil will be permanent. If we hope to avoid economic depression and social disintegration, we must act immediately to conserve energy and find substitutes for those forms of energy that are about to become scarce.

Energy Is Abundant and Will Remain So

Many people believe that the current energy crisis, like those of the past, is temporary. Stephen Moore, a senior fellow at the Cato Institute, makes the optimists' case:

> The human intellect is the most valuable resource on this Earth, and thanks to our ingenuity, we are always finding new supplies and kinds of resources on the Earth. For example, thanks to leaps forward in technology, we can drill deeper into the Earth's crust to recover and use oil and other valuable resources. Technological advances allow us to find substitutes for resources that may genuinely be running low in supply.[1]

Moore also reminds us that, in the past, doomsayers warned that we faced dire energy shortages but were proven wrong every time.

There is enough oil to last for decades.

Figures in the *BP Statistical Review of World Energy*, which are considered authoritative, contradict the pessimists. In 2006, BP estimated the world's proven oil reserves at 1.21 trillion barrels, enough to last 40 years at the current rate of consumption. The BP figures suggest that we have ample time to plan for oil depletion. Furthermore, reserves have increased over the years—in 1986, they stood at only 877 billion barrels—as new oil fields have been discovered. As a result, the reserves-to-production ratio is the same today as it was 20 years ago.

Robert Bradley, the president of the Institute for Energy Research, notes that the world's oil reserves today are more than 15 times greater than they were when recordkeeping began in 1948. David Deming, a professor at the University of Oklahoma, goes on to say that speaking about "reserves" is misleading because it creates the impression that oil will run out much sooner than it actually will. Deming compares oil reserves to food in a family's pantry. Dividing what is in the pantry by the family's daily food consumption would lead to the conclusion that the family will starve to death in a few weeks. That never happens, however, because the family keeps restocking the pantry.

Exploring for and recovering oil is the equivalent of restocking the pantry. Had we not continually replenished our supply, Deming argues, America would have run out of oil by 1930. The amount of worldwide oil reserves has been increasing because technology has made it possible to find new sources of oil. M.A. Adelman, a professor at the Massachusetts Institute of Technology (MIT), explains:

> Growing knowledge lowers cost, unlocks new deposits in existing areas, and opens new areas for discovery. In 1950, there was no offshore oil production; it was highly "unconventional" oil. Some 25 years later, offshore wells were being drilled in water 1,000 feet deep. And 25 years after that, oilmen were drilling in water 10,000 feet deep—once technological advancement

enabled them to drill without the costly steel structure that had earlier made deep-water drilling too expensive.[2]

It appears that there is much more oil waiting to be discovered. In 2007, the Government Accountability Office (GAO) reported to Congress on the prospect of peak oil. The report cited a 2000 estimate by the U.S. Geological Survey that the world has as much as 732 billion barrels of undiscovered conventional oil,

Figures from the *BP Statistical Review of World Energy*

For more than 50 years, the *BP Statistical Review of World Energy** has been a widely respected and authoritative publication in the field of energy research. BP (formerly known as British Petroleum), which compiles the report, is one of the world's largest oil and gas companies.

Here are some of the key findings from the most recent report, released in 2007. Unless otherwise indicated, production and consumption figures are from 2006:

Oil:

- The world's proven oil reserves were 1.21 trillion barrels, up from 877.4 billion barrels in 1986.
- Worldwide oil production was 81.7 million barrels per day, up from 69.9 million barrels per day in 1996.
- Worldwide oil consumption was 83.7 million barrels per day, up from 71.5 million barrels per day in 1996. In 2006, consumption rose by only 0.7 percent, the weakest growth since 2001 and half the 10-year average.
- The United States was the largest consumer of oil, accounting for 24.1 percent of world oil consumption, followed by the 25 European Union countries (18.2 percent), China (9 percent), and India (3.1 percent).
- Output from OPEC countries rose by 130,000 barrels per day. Output from non-OPEC countries rose by about 600,000 barrels per day. Russia increased its production by 220,000 barrels per day, while Azerbaijan,

and that one-quarter of it could be found in the Arctic. The GAO added: "However, relatively little exploration has been done in this region, and there are large portions of the world where the potential for oil production exists, but where exploration has not been done."[3]

Meanwhile, on the demand side, buildings and machinery are becoming increasingly energy-efficient: Every year, it takes less energy to do the same amount of useful work. At the same

Angola, and Canada each increased production by 100,000 barrels per day.

Natural gas:

- Production increased by 3 percent, slightly above the 10-year average. Russia accounted for the largest incremental growth.
- Worldwide consumption grew by 2.5 percent, less than in 2005 but close to the 10-year average.

Coal:

- In terms of consumption, coal remained the world's fastest-growing fossil fuel, which implies that carbon dioxide emissions will continue to increase.
- Worldwide consumption grew by 4.5 percent, well above the 10-year average. China's consumption rose by 8.7 percent, accounting for more than 70 percent of the worldwide increase.

Other findings:

- Nuclear power output increased by 1.4 percent.
- Hydroelectric power generation increased by 3.2 percent.
- High prices for fossil fuels, combined with renewed government interest, have encouraged the rapid growth of renewable energy sources. This growth, however, starts from a very small base.

*Source: BP PLC. *BP Statistical Review of World Energy June 2007*. London, England, 2007.

time, our automakers are developing more fuel-efficient versions of the internal-combustion engine. For example, the 1992 Honda Civic VX hatchback got 55 miles per gallon on the highway, compared with 35 mpg for the 1991 model. The added fuel efficiency was the result of more advanced engine design. The 1999 Honda Insight, which weighed slightly less than the Civic, got 70 mpg on the highway, representing a doubling of gas mileage in Honda's smallest car in less than a decade.

If oil becomes scarce, substitutes are available.

Energy optimists are fond of saying "the Stone Age did not end for lack of stones," which means that humans turn to new energy sources long before existing ones run out. Deming elaborates:

> No technology since the birth of civilization has been sustainable. All have been replaced as people devised better and more efficient technologies. The history of energy use is largely one of substitution. In the 19th century, the world's primary energy source was wood. Around 1890, wood was replaced by coal. Coal remained the world's largest source of energy until the 1960s when it was replaced by oil. We have only just entered the petroleum age.[4]

If oil starts to run out, there are adequate supplies of conventional fuels—coal, natural gas, and nuclear power—that we can use in its place.

At one time, natural gas was considered a nuisance and either left in the ground or burned off. After a network of pipelines was built following World War II, people used gas to heat businesses and homes and to operate appliances. Paul Roberts explains the advantages of this fuel:

> Gas is now the abundant fuel: by some estimates, the massive gas fields in Qatar, Iran, Turkmenistan, and Russia,

> which hold more than half the known global reserves, could fuel the world for more than half a century, and these are only a fraction of total gas assets. Gas is also incredibly versatile. It can be used in everything from power plants to gas-powered buses and taxis. It can be converted to liquid fuels—gasoline, for example—and compete directly with oil.[5]

Roberts adds that natural gas emits less pollution and fewer greenhouse gases, and can be easily refined into pure hydrogen to power fuel cells and other technologies of the future.

America is also rich in coal. In 2003, according to the National Mining Association, our reserves totaled 496 billion tons, enough to last for more than 200 years. The utility industry, which relies on coal to produce about half the nation's electric power, is making power plants more efficient. That will cut down on emissions as well as the amount of coal used. Scientists are also developing "clean coal" technology, which chemically "washes" pollutants out of burning coal, and exploring ways of "sequestering" carbon dioxide: storing it underground rather than letting it escape into the atmosphere. Nuclear energy is another potential source. Currently, about one-fifth of America's electric power is generated at nuclear plants, even though construction of new plants came to a halt after the incident at the Three Mile Island plant in 1979. Nuclear power produces less pollution than coal or natural gas, and we do not have to rely on foreign sources for uranium, the element whose atoms are split in the process of generating power.

In addition, we have barely begun to exploit so-called "unconventional" oil, which is more difficult to recover and turn into usable fuel. "Heavy oil," which does not flow easily, and oil sands, which contain very heavy oil, represent the largest sources of unconventional oil. Canada is a leading source of oil sands.

(continues on page 50)

When Will Peak Oil Arrive?

In 2007, the National Energy Technology Laboratory (NETL) issued a report* summarizing experts' forecasts about the peaking of oil production.

Before citing the forecasts, the NETL examined the problem of supply. According to the International Energy Agency (IEA), the discovery of new oil has fallen sharply since the 1960s, and in the past decade, discoveries have replaced only half the oil produced. The decline in oil discoveries has been most dramatic in the Middle East, where they fell from 187 billion barrels in 1963–1972 to 16 billion barrels in 1993–2002. According to a 2005 study by the Royal Swedish Academy of Sciences, production was declining in 54 of the 65 most important oil-producing countries, and the rate of discovery of new reserves was less than one-third of the present rate of consumption.

One important variable is the reliability of figures given by countries that belong to OPEC. The NETL observed: "A number of forecasters have accepted OPEC reserves estimates at face value in part because there is no independent source of verification. This acceptance is troubling in light of the fact that past history raises significant questions about the validity of OPEC reporting."

Despite differing estimates of the size of the world's oil reserves, the NETL stated that these differences would not significantly change the date when production peaked. Assuming that demand will grow by 2 percent a year, even the addition of 900 billion barrels to the estimated reserves, which represents more oil than has been produced to date, would delay peaking by only 10 years.

Not only is less oil being discovered, but there are also problems in recovering the oil that exists. One problem is a lack of investment. Both the IEA and a panel sponsored by the Aspen Institute found that worldwide investment was not sufficient to maintain production. Thus, lack of capital and skilled people, not a lack of oil, might bring on the peaking of production.

Turning to peak oil forecasts, NETL cited the following experts who believe that oil production either has peaked or will peak within the next five years (by 2012): oil and gas investor T. Boone Pickens (says production peaked in 2005); Kenneth Deffeyes, a professor emeritus at Princeton University who was a geologist at Shell Oil (says the peak occurred in December 2005); E.T. Westervelt and his colleagues at the Army Corps of Engineers (says the peak is "at hand"); Ali Samsam Bakhtiari, a planner at the Iranian National Oil Company (says the peak is "now"); Roger Herrera, a retired geologist at BP (says we are close to or past the peak); Henry Groppe, an oil and gas expert and businessman ("very soon"); Stephen Wrobel, an investment fund manager (by 2010); Dr. Roger Bentley, a university energy analyst (around 2010); Colin Campbell, a retired oil company geologist (2010); Chris Skrebowski, the editor of *Petroleum Review* (says 2010, plus or minus a year); and Leif Meling, a geologist at Norway's Statoil (says around 2011).

Those experts who believe that production will peak within 5 to 15 years (between 2012 and 2022) include: Pang Xiongqi and his colleagues at the China University of Petroleum (around 2012); Rembrandt Koppelaar, a Dutch oil analyst (around 2012); the staff of Volvo Trucks (by 2017); Christophe de Margerie, an oil company executive (by 2017); Sadad al Husseini, a retired executive vice president of Saudi Aramco (2015); the brokerage company Merrill Lynch (around 2015); J. Robinson West, of the consulting firm PFC Energy (2015 to 2020); Charles T. Maxwell, of the brokerage company Weeden & Company (around 2020 or earlier); and the French oil company Total (around 2020).

Those who believe that production will not peak for at least 15 years (2022 or later) include the financial services company UBS (mid- to late 2020s) and Cambridge Energy Research Associates (CERA) (well after 2030). The oil company ExxonMobil argues that there is no sign of peaking; and John Browne, the head of the oil company BP, argues that it is impossible to predict when the peak will come. OPEC has denied peak oil theory, and Mark Morrison of CERA has called the theory "garbage."

The Energy Information Agency (EIA) is the statistics-gathering arm of the U.S. Department of Energy. In 2000, the EIA concluded that world oil production would peak in 2016. Four years later, the EIA reconsidered its earlier work and concluded: "In any event, the world production peak for conventionally reservoired crude is unlikely to be 'right around the corner' as so many other estimators have been predicting. Our analysis shows that it will be closer to the middle of the 21st century than to its beginning." More recently, EIA verified that it still does not forecast oil peaking before 2030.

*National Energy Technology Laboratory, *Peaking of World Oil Production: Recent Forecasts.* Washington, DC: U.S. Department of Energy, 2007.

(continued from page 47)
Oil shale, a mixture of chemicals that can be converted into fuel, is another possibility. Much of it is believed to be in the western United States. There are potentially vast reserves of unconventional oil; in fact, the International Energy Agency estimates that they may total 7 trillion barrels. Unconventional oil accounts for only 3 percent of the world's liquid fuel supply. As conventional oil supplies diminish, however, unconventional oil could account for a much larger percentage.

Energy pessimists have been proven wrong.

Adelman argues that dire predictions about energy have repeatedly been proven wrong:

> In 1875, John Strong Newberry, the chief geologist of the state of Ohio, predicted that the supply of oil would soon run out. The alarm has been sounded repeatedly in the many decades since. In 1973, State Department analyst James Akins, then the chief U.S. policymaker on oil, published [in *Foreign Affairs* magazine] "The Oil Crisis: This time the wolf is here," in which he called for more domestic production and for improved relations with oil-producing nations in the Middle East. In 1979, President Jimmy Carter, echoing a CIA assessment, said that oil wells "were drying up all over the world." . . . The doomsday predictions have all proved false. In 2003, world oil production was 4,400 times greater than it was in Newberry's day, but the price per unit was probably lower. Oil reserves and production even outside the Middle East are greater today than they were when Akins claimed the wolf was here. World output of oil is up a quarter since Carter's "drying up" pronouncement, but Middle East exports peaked in 1976–77.[6]

Some experts, such as Deming, dispute M. King Hubbert's prediction of oil depletion. Deming points out that Hubbert underestimated oil production in this country: "Production in the 48 contiguous states peaked, but at much higher levels than Hubbert predicted. From about 1975 through 1995, Hubbert's upper curve was a fairly good match to actual U.S. production data. But in recent years, U.S. crude oil production has been consistently higher than Hubbert considered possible."[7] Deming adds that, in 2000, oil production in the lower 48 states was 1.7 times higher than what Hubbert predicted in 1980. Some experts also contend that the Hubbert curve itself is misleading. Adelman notes that coal production in Europe peaked in 1913 and is insignificant today. Peaking occurred, however, because it no longer made economic sense to mine coal, not because the supply ran out. There are still millions of tons in the ground in Europe.

Even today, energy is plentiful and cheap.

Many experts contend that high energy prices do not necessarily mean that the supply is running out. In the Middle East, the world's primary source of oil, the cost of recovering oil remains low. Peter Huber and Mark Mills explain:

> The cost of oil comes down to the cost of finding, and then lifting or extracting. First, you have to decide where to dig. Exploration costs currently run under $3 per barrel in much of the Mideast, and below $7 for oil hidden deep under the ocean. But these costs have been falling, not rising, because imaging technology that lets geologists peer through miles of water and rock improves faster than supplies recede. Many lower-grade deposits require no new looking at all.[8]

Price-fixing helps keep oil prices much higher than the actual cost of recovering it. The 12 countries that belong to OPEC control more than one-third of the world's oil production, giving them considerable leverage over prices. Acting in concert, OPEC members can hold back production and prop up the price of oil, or flood the market with oil to drive down the price and put would-be competitors out of business. Huber and Mills explain:

> Investing billions in tar-sand refineries is risky not because getting oil out of Alberta is especially difficult or expensive, but because getting oil out of Arabia is so easy and cheap. . . . Investing $5 billion over five years to build a new tar-sand refinery in Alberta is indeed risky when a second cousin of Osama bin Laden can knock $20 off the price of oil with an idle wave of his hand on any given day in Riyadh.[9]

The political situation in oil-producing countries also affects prices. Vijay Vaitheeswaran explains: "Petroleum is probably the only global business in which the industry's largest firms and best assets are controlled by governments. It may seem astonishing, but even the likes of Chevron and Texaco are midgets compared with the state-run oil giants like Saudi Arabia's Aramco. The industry's lowest-cost reserves are also controlled by governments (think Saudi Arabia and Iraq, for a start)."[10] State-owned companies, which do not answer to shareholders, tend to be inefficient. They are often run by political appointees or operate in a culture where corruption is a way of life. Many countries bar foreigners from investing in their oil industry, a policy that closes them off to capital and technology that would allow them to recover oil more efficiently. Others allow foreign investment but are bad places to do business. In the words of Huber and Mills, "Oil prices gyrate and occasionally spike—both up and down—not because oil is scarce, but because it's so abundant in places where good government is scarce."[11]

The price of energy is also affected by the quality of the infrastructure—that is, the network of pipelines, power grids, and refineries that produce and transport energy. In 2001, the National Energy Policy Development Group said that obsolete infrastructure was more to blame than lack of supply for the United States's energy problems:

> Our energy infrastructure has failed to keep pace with the changing requirements of our energy system. Domestic refining capacity has not matched increases in demand, requiring the United States to import refined products. Natural gas pipelines have not expanded sufficiently to meet demand. The electricity transmission system is constrained by insufficient capacity.[12]

At any rate, energy remains relatively cheap, even at today's prices. The price of a kilowatt-hour of electricity has steadily fallen since Thomas Edison's first power station opened in New York in 1882. Until the 2008 price spike, a barrel of oil cost less on an inflation-adjusted basis than it did during the lifetime of Colonel Drake, who drilled the nation's first oil well. In addition, household budgets are less vulnerable to high energy prices than they were in the 1970s. The authors of the Hirsch Report noted that Americans' per-capita incomes doubled between 1981 and 2005, and that energy had become a much smaller component of expenditures. For example, in 1985, when oil prices were low, gasoline and oil represented 20 percent of the cost of owning and operating a vehicle. By 2002, that figure had fallen to 10 percent, in part because of improved fuel efficiency. Furthermore, the American economy is less dependent on oil than it was during the 1970s. That is because less energy-intensive sectors, such as banking and retail, have replaced manufacturing. In fact, energy was so cheap for so long, our economy has tolerated serious inefficiencies in the process of generating and transporting it.

(continues on page 56)

FROM THE BENCH

Center for Biological Diversity v. NHTSA (2007)

After the OPEC oil embargo, Congress enacted the Energy Policy and Conservation Act (EPCA) of 1975 (Public Law 94–163). That legislation imposed Corporate Average Fuel Economy (CAFE) standards on automakers. Beginning with the 1978 model year, passenger automobiles had to average 18 mpg. The current standard is 27.5 mpg and will rise to 37.5 mpg by 2020.

Congress left it to the National Highway Traffic Safety Administration (NHTSA) to set CAFE standards for vehicles other than passenger cars. At the time, cars and trucks were distinct classes of vehicles. Trucks were expected to consume more gasoline because they were designed for heavy-duty uses. Few Americans drove trucks for pleasure or to commute to and from work. Accordingly, NHTSA set lower mpg standards for "light trucks," which it defined as vehicles designed for off-highway operation or designed to perform at least one of five specific functions, such as greater cargo-carrying volume or a flat floor to haul items.

In 2001, Congress directed the National Academy of Sciences to study the effectiveness of CAFE standards. One of the academy's findings was that lower CAFE standards for light trucks encouraged automakers to build minivans and sport-utility vehicles and promote them to consumers as alternatives to large sedans and station wagons. As a result, consumers were buying as many light trucks as passenger cars, and the fuel economy of America's car and light-truck fleet had fallen from 25.9 mpg in 1987 to 24 mpg in 2000.

Afterward, NHTSA began a rulemaking process aimed at setting new CAFE standards for light trucks. In August 2005, the agency proposed a stepwise increase to 23.5 mpg by the 2010 model year and, after that, a miles-per-gallon standard based on a vehicle's "footprint," its length times its width. In setting those standards, NHTSA relied heavily on a cost-benefit analysis that concluded that higher standards would result in "adverse economic consequences, such as a significant loss of jobs or the unreasonable elimination of consumer choice."

Some states and cities, along with several public-interest organizations, went to court to force NHTSA to adopt higher CAFE standards. (For example, the advocacy group Environmental Defense recommended a 26-mpg standard by the 2011 model year.) They filed a petition in the U.S. Court of Appeals for the Ninth Circuit.

There they argued that NHTSA violated the EPCA by emphasizing the standards' economic effect on the auto industry at the expense of energy conservation, and also the National Environmental Policy Act, by not considering the impact of CAFE standards on global warming.

In *Center for Biological Diversity v. National Highway Traffic Safety Administration*, 508 F.3d 550 (9th Cir. 2007), a three-judge panel concluded that NHTSA had broken the law. Judge Betty Fletcher wrote the court's opinion. She first concluded that NHTSA had violated the EPCA. Even though that law allowed NHTSA to use a cost-benefit analysis, it did not allow the agency to set fuel-economy standards that were contrary to its goal of conservation.

The court concluded that NHTSA also violated the law by failing to "monetize," or calculate the economic benefits of, reducing greenhouse-gas emissions. Fletcher said: "NHTSA fails to include in its analysis the benefit of carbon emissions reduction in either quantitative or qualitative form. It did, however, include an analysis of the employment and sales impacts of more stringent standards on manufacturers."

The court rejected NHTSA's claim that if it monetized the benefits of reducing emissions, it also would have to monetize the added safety costs resulting from Americans driving lighter vehicles on account of higher CAFE standards. Fletcher went on to say: "Consumers use light trucks primarily for passenger-carrying purposes in large part because that is precisely the purpose for which manufacturers have manufactured and marketed them." She observed that the National Academy of Sciences had recommended narrowing the definition of light truck, a step the Environmental Protection Agency had already taken in setting emissions standards.

In addition, the court concluded that NHTSA's assessment of the new CAFE standards' environmental consequences was not thorough enough and ordered the agency to prepare a new assessment. In doing so, the judge rejected the argument that the EPCA, which governed fuel economy, exempted the agency from considering environmental impacts such as global warming in setting CAFE standards. Fletcher wrote: "Energy conservation and environmental protection are not coextensive, but they often overlap." She also cited "the compelling scientific evidence" that linked carbon dioxide and other heat-trapping gases to global warming.

(continued from page 53)

Summary

On a number of occasions, those who predicted that we would soon run out of energy turned out to be wrong. Eventually, peak oil pessimists will be proven wrong as well. There is enough conventional oil to meet the world's needs for decades to come, and there are potentially vast reserves of "nonconventional" oil that can be recovered in the future. Even if oil production begins to decline, we can rely on conventional fuels such as natural gas, nuclear energy, and coal for years without having to rely on untested alternatives. The 2008 energy crisis did not result from a lack of supply so much as from factors such as an obsolete energy infrastructure and price manipulation by oil-producing countries. In any event, energy remains cheap, and its current high price has less of an effect on the economy than it had during the 1970s.

Continued Reliance on Fossil Fuels Is Dangerous

Professor Michael Klare describes a little-known but significant event in American history:

> On February 14, 1945, President Franklin D. Roosevelt met with Ibn Saud, the Saudi king, aboard a Navy ship. Even though there is no record of what the two men talked about, most historians and government officials think the two leaders forged a tacit alliance—one which obliged the United States to protect Saudi sovereignty and independence in return for a Saudi pledge to uphold the American firms' dominance of the oil fields. Whether or not this was really the case, leaders of both countries have acted as if it were.[1]

Today, Saudi Arabia is the leading foreign supplier of oil to this country. Experts warn that our dependence on oil from Saudi

Arabia and other Middle Eastern countries has entangled us in the politics of that region and made us vulnerable to economic rivals, hostile regimes, and even terrorists.

Undependable governments supply us with energy.

According to the *BP Statistical Review of World Energy*, the United States consumes 24.1 percent of the world's oil but accounts for only 8 percent of production. This huge disparity is made up by oil imports. Our oil supply, however, can be disrupted by the governments of oil-producing countries. Some are openly hostile to us. They include Iran, which President

"NOPEC": Congress Addresses OPEC Price-Fixing

The Organization of the Petroleum Exporting Countries (OPEC) was formed in 1960 to give oil-producing countries more leverage in their negotiations with oil companies. Over the years, OPEC's power has grown considerably. Today, it can drive up the world price of oil whenever its members agree to limit production. Thus, OPEC is a classic example of a *cartel*, a group of entities that act together to affect market prices by controlling the production and marketing of a product.

OPEC-type cartels have been illegal in the United States since 1890. That year, Congress passed the Sherman Antitrust Act (Title 15, Section 1 and following of the U.S. Code), which provides: "Every contract, combination in the form of trust or otherwise, or conspiracy, in restraint of trade or commerce among the several States, or with foreign nations, is declared to be illegal." Antitrust law violators can face criminal charges as well as civil lawsuits brought by victims of their illegal actions. A victim who wins a civil suit is awarded three times the amount of his or her damages.

The Sherman Act was directed at American companies such as John D. Rockefeller's Standard Oil Company. Its language does not rule out suits against foreign countries, but those suits are restricted by "act of state" doctrine, under which a country's actions within its own borders cannot be questioned in another country's

George W. Bush called part of the "axis of evil"; and Venezuela, which sells oil to Cuba at below-market prices and uses oil revenue to finance left-wing activists and even insurgent groups elsewhere in Latin America.

For a variety of reasons, other oil-producing countries are undependable. Klare explains: "Ethnic violence has hampered output in Nigeria.... Insurgent warfare has crippled Colombia. Georgia, a major transit state for the BTC [Baku-Tblisi-Caspian] pipeline, faces an array of separatist conflicts. Diversify its imports as it may, the United States will not be able to escape the violence and disorder that seems to follow the oil industry wherever it goes."[2] *Oil & Gas Journal* recently estimated that

courts. The act of state doctrine is not part of international law, but federal courts in this country apply it in order to avoid interfering with the president's constitutional authority to conduct relations with foreign countries.

In *International Association of Machinists and Aerospace Workers v. OPEC*, 649 F.2d 1354 (9th Cir. 1981), a labor union attempted to sue OPEC. A federal appeals court held that the act of state doctrine prevented it from hearing the merits of that lawsuit. Some legal experts and members of Congress believed that the appeals court came to the wrong conclusion. They argued that the act of state doctrine does not apply to countries engaged in purely economic activity.

In 2007, as the price of oil rose, the House of Representatives overwhelmingly passed the "No Oil Producing and Exporting Cartels Act of 2007" or "NOPEC" Act (H.R. 2264). Section 2 of NOPEC would extend the Sherman Act to foreign countries acting as a cartel to limit the production or distribution of, fix the price of, or otherwise restrain trade in the American market for petroleum products. NOPEC also "overrules" the decision in the *International Association of Machinists* case by specifically making the act of state doctrine inapplicable to lawsuits against oil cartels. The attorney general may file a lawsuit against OPEC, but private citizens may not sue on their own.

The NOPEC Act died in the Senate. It is not clear if the 111th Congress, which started work in January 2009, will pass similar legislation or if President Obama will sign it.

more than 60 percent of world oil reserves are found in countries where unstable political conditions could limit exploration and production.

At the same time, much of the world's oil supply is now controlled by government officials rather than oil company executives. Joshua Kurlantzick, a visiting scholar at the Carnegie Endowment, explains: "These days, Western multinationals control just 10 percent of the world's oil, while state-run firms, according to a November 2007 paper from Rice University's James A. Baker III Institute for Public Policy, exercise exclusive domain over roughly 77 percent."[3] State-owned companies do not have to show a profit, so government officials can use them to achieve political and diplomatic goals. One notable example may be found in China. PetroChina, the state-run oil company, is the world's most valuable company, and its parent firm, China National Petroleum Corporation, is looking for oil around the world. Matthew Simmons, a banker who has written extensively about oil depletion, observes: "China has forged agreements with three of the largest petroleum exporters—Saudi Arabia, Iran, and Venezuela—and with several others. Not surprisingly, several of these exporting countries are currently in disputes with the United States. These countries may not be above using their increased market leverage in ways that will damage U.S. interests."[4]

Relying on foreign oil invites terrorism.

Some people believe that the terrorist attacks of September 11, 2001, were the result of U.S. dependence on foreign oil. Fifteen of the 19 men who carried out those attacks were citizens of Saudi Arabia, whose government in effect subsidized their radical beliefs with oil money. Paul Roberts explains: "Desperate to curb fundamentalist opposition at home, Saudi royals gave huge bribes to radical mosques, then financed the Islamic 'revolution' in places like Afghanistan and Pakistan, essentially exporting a generation of young Saudi radicals and sowing the seeds of today's militant Islam."[5] Someday, those radicals might carry out an attack that causes more economic disruption than the September 11 attacks did. According to author Lutz Kleveman, "There is a growing risk that radical Islamist groups will topple the corrupt Saud dynasty, only to then stop the flow of oil to 'infidels.' The consequences of 8 million barrels of oil—10 percent of global production—disappearing from the world markets overnight would be disastrous."[6]

America's dependence on foreign oil requires us to send troops all over the world to maintain access to it. Our stepped-up presence overseas, however, invites backlash. Many believe that Osama bin Laden, who masterminded the September 11 attacks, was motivated by the presence of American troops on Saudi soil. Klare warns that our foreign policy will breed more terrorists such as bin Laden:

> The growing American reliance on unstable suppliers in dangerous parts of the developing world is creating social, economic, and political pressures that are exacerbating local schisms and so *increasing* the risk of turmoil and conflict. This is the case for a number of reasons. First, the conspicuous presence of U.S. oil firms is bound to arouse hostility from people who reject American values or resent the great concentration of wealth and power in America's hands. Second, the very production of oil in otherwise underdeveloped societies often skews the local economy—funneling vast wealth to a few and thus intensifying the preexisting antagonism between the haves and the have-nots.[7]

Finally, our energy infrastructure itself invites terrorist attacks. Shortly after September 11, the Union of Concerned Scientists (UCS) issued a report that warned that continued dependence on fossil fuels endangered our national security. It criticized the Bush administration's reaction to the attacks, which emphasized producing more oil and gas at home rather than conservation and alternative energy. The UCS contended that the administration's approach would make us less secure. Not only would this approach leave our economy more susceptible to price shocks, it would lead to the construction of large installations that are high on the list of terrorist targets. The targets include refineries and oil and gas pipelines, which use traditional sources of energy.

We face the prospect of wars over oil.

Our military presence in oil-rich regions increases the chances that we will have to fight a war over that oil. For years, the United States has warned the rest of the world that it would use force to

Was Iraq a War for Oil?

In 2003, a coalition of countries led by the United States invaded Iraq and drove that country's dictator, Saddam Hussein, from power. President Bush argued that the invasion was necessary for several reasons, the most important of which was that Saddam was developing weapons of mass destruction—chemical, biological, and even nuclear weapons.

Some critics, however, maintain that the real motivation for the war was access to Iraq's large reserves of oil. Michael Klare, a professor at Hampshire College in Massachusetts, wrote in 2004: "Security was directly tied to the safety of Persian Gulf oil supplies and thus to the prospects for increased output. So long as Saddam Hussein remained in power, the Gulf would never be entirely stable, and the United States would never be able to boost Iraqi petroleum production."* Klare added that, both before and after the Iraq war, the United States increased its military presence in the oil-producing regions of the Middle East and central Asia.

Paul Rogers, a professor of peace studies at Bradford University in England, went even farther in his criticism: "Whether or not direct lobbying had an effect, the result of the Iraq war has certainly been 'good' for the oil industry. Oil-related companies like Halliburton have found Iraq immensely profitable, and the oil companies themselves are currently enjoying exceptional profitability."** He discounted the Bush administration's contention that invading Iraq would promote democracy in the region, saying that such talk is "only relevant if it ensures increasing US protect its access to Middle East oil. Paul Roberts calls the 1991 Persian Gulf War "the first military conflict in world history that was entirely about oil." He goes on to say: "Interest in oil had not only prompted Saddam [Hussein] to invade but had largely defined the world's reaction. While government spokesmen made much of Kuwaiti suffering, behind closed doors, diplomats were focused almost entirely on the loss of Kuwaiti oil and, more to the point, whether the attendant price spike would tip the world into recession."[8]

Likewise, some observers believe that the American-led invasion of Iraq in 2003 was motivated by access to the region's

influence, backed by powerful military forces." Rogers also predicted that, before long, foreign-policy experts would agree with him that the war was "essentially motivated by oil."

Charles Kohlhaas, a former professor of petroleum engineering at the Colorado School of Mines, rejects "war for oil" claims. Days before the war started, Kohlhaas argued that it would make no economic sense to invade Iraq. He said: "The most common concern regarding the possible effect of an invasion on oil production is that oil operations will be disrupted during military action. Disruption probably will reduce world supplies and drive oil prices up on the world markets for a short-term.... The most likely outcome of an Iraqi invasion is a reduction of supplies and increased prices; clearly an additional cost attributable to an invasion, not a benefit, and exactly contrary to a claim that the invasion is 'for the oil.'" He went on to say: "As a business decision, invading Iraq 'for the oil' is a loser, a big loser. Anyone who would propose, in a corporate boardroom, invading Iraq for the oil would probably find his career rather short. No, the slogan 'no war for oil' is a blatant misrepresentation propagated for political reasons."***

* Michael T. Klare, *Blood for Oil: The Dangers and Consequences of America's Growing Dependency on Imported Petroleum*. New York: Metropolitan Books, 2004, p. 98.

** Paul Rogers, "It's the Oil, Stupid," www.opendemocracy.net, March 24, 2005. http://www.opendemocracy.net/conflict/article_2393.jsp.

*** Charles A. Kohlhaas, "War in Iraq: Not a 'War for Oil,'" *In the National Interest*. 9, no. 2 (March 5, 2003). http://www.inthenationalinterest.com/Articles/Vol2Issue9/vol2issue9kohlhaas.html.

oil supply. Klare predicts the war in Iraq will not be the last one fought in that region: "Although we cannot foresee the precise nature and timing of the next crisis the [U.S.] Central Command will address, it's safe to predict that its forces will see combat in the Persian Gulf once more—and that such intervention will be repeated again and again until the last barrel of oil is extracted from the Gulf's prolific but highly vulnerable reservoirs."[9]

Looking farther ahead, some warn that the world's superpowers could go to war over oil. China, which has a huge population and a fast-growing economy, is of particular concern. Klare explains:

> For the United States, China will become a major rival for new oil, and its pursuit of a larger share of the region's output may contribute to tighter supplies and higher prices. Like its rivals, Beijing will no doubt cultivate closer economic and political ties with local producers—if need be, by stepping up its transfers of arms and military technology. This will inevitably accelerate local arms races and exacerbate regional tensions; it could even lead to conflict with Washington if the recipients are hostile regions, like Iran.[10]

Meanwhile, there are signs that Russia is behaving like the old Soviet Union. The Washington-based monitoring group Freedom House downgraded that country from "partly free" in 2002 to "not free" in 2007. Recently, Russian Prime Minister Vladimir Putin sent troops into the former Soviet republic of Georgia in response to ethnic fighting there, and has threatened to cut off natural gas shipments to neighboring countries. Although it has its own oil resources, Russia, like China, is close to the oil resources of Central Asia and the Middle East and could use force to protect its access—or deny it to other countries. For that reason, some observers warn that those regions could become a "global chessboard," with the United States, Russia, and China embroiled in a three-way conflict for the world's remaining oil.

Oil revenue funds dictatorships and breeds conflict.

High oil prices during the 1970s resulted in an outflow of money from oil-consuming nations to states that were unprepared to handle their newfound wealth. According to Michael Ross, a professor at the University of California, Los Angeles, these countries became victims of the "oil curse":

> The affliction hits when a country becomes a significant producer and exporter of natural resources. Rising resource exports push up the value of the country's currency, which makes its other exports, such as manufactured and agricultural goods, less competitive abroad. Export figures for those products then decline, depriving the country of the benefits of dynamic manufacturing and agricultural bases and leaving it dependent on its resource sector and so at the mercy of often volatile international markets.[11]

Ross argues that oil money turns countries into dictatorships and keeps them in power: "The world's thirst for oil immunizes petroleum-rich governments from the kind of pressures that might otherwise force them to the bargaining table. Since these governments' coffers are already overflowing, aid means little to them. They can readily buy friends in powerful places and therefore have little fear of sanctions from the UN Security Council."[12] At home, dictatorships spend oil revenue to enrich their political allies, and on police and security forces to keep themselves in power. For the most part, they ignore the needs of their poorest citizens: Infant survival, nutrition, life expectancy, literacy, and education levels are worse in oil-producing countries than in non-oil-producing countries.

Many oil-producing countries have a large disparity between rich and poor and few opportunities for peaceful protest. These countries are at risk of violent insurgency, even civil war. Even though the world is more peaceful now than it was at the end of the Cold War, that is not true of oil-producing countries. According to Ross, those countries are home to one-third of the world's civil wars and are twice as likely to suffer from internal rebellion, in part because oil money funds rebel groups. Insurgents can steal oil and sell it on the black market, as has happened in Iraq and Nigeria; extort money from oil companies working in remote areas, as in Colombia and Sudan; or find business partners willing to fund them in exchange for future considerations in the event they seize power, as in Equatorial Guinea and the Republic of the Congo.

Insurgency feeds a vicious circle. The authors of the Hirsch Report said: "The factors that cause oil price escalation and volatility could be further exacerbated by terrorism. For example, in the summer of 2004, it was estimated that the threat of terrorism had added a premium of 25–33 percent to the price of a barrel of oil. As world oil peaking is approached, it is not difficult to imagine that the terrorism premium could increase even more."[13] Higher oil prices, in turn, make oil infrastructure such as pipelines and refineries even more attractive to terrorists. In Nigeria, for example, insurgent activity has reduced that country's production, contributing to the increase in the world price of oil in recent years. Graham Bowley of the *New York Times*, who covered that conflict, observed that so long as oil prices remain high, the rebels recognize the power they have; and that the rebels, in turn, are one reason prices are likely to remain high.

Fossil fuels damage the environment.

Our continued dependence on fossil fuels also threatens the future of the planet. When these fuels are burned, they release carbon dioxide and other so-called "greenhouse gases" that trap solar radiation, causing the Earth's temperature to rise. The

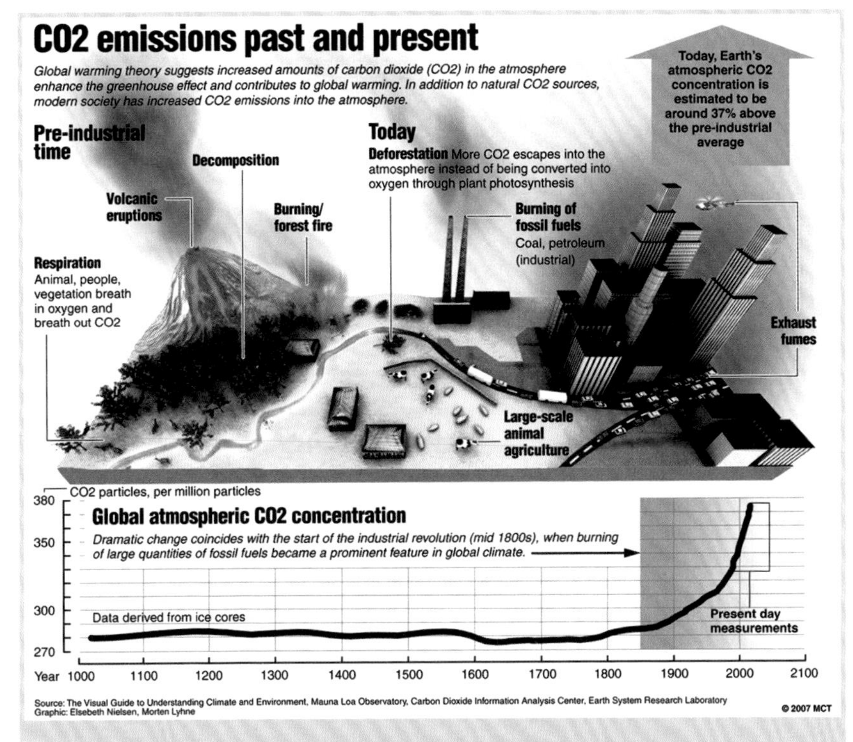

The top graphic shows the natural and manmade ways CO_2 is released into the atmosphere; below it, the significant increase of CO_2 in the atmosphere since the industrial revolution of the nineteenth century is depicted.

Intergovernmental Panel on Climate Change (IPCC) concluded: "Warming of the climate system is unequivocal, as is now evident from observations of increases in global average air and ocean temperatures, widespread melting of snow and ice and rising global average sea level."[14] The IPCC also said that there is at least a 90 percent chance that human beings are responsible for global warming. Unless greenhouse-gas emissions are substantially reduced, the IPCC warns, the world risks much larger temperature increases, which will likely produce severe droughts and flooding, widespread famine, conflict over water and other resources, and even a massive die-off of plant and animal species.

Energy depletion and global warming not only reinforce each other, but shortages of oil could aggravate the global-warming crisis. Recently, the Oil Depletion Analysis Centre reported: "A common misconception is that 'running out of oil' would at least be good for climate change. Unfortunately the opposite may be true. The 'alternative' transport fuels advocated by climate change campaigners are likely to be inadequate, and the kinds of replacement fuels which might fill some of the yawning post-peak fuel deficit are likely to worsen global warming."[15] The problem is worse yet in the developing world. China, for example, gives economic growth a higher priority than protecting the environment. By some estimates, its government is building one coal-fired electric power plant a week, and many of these plants use obsolete, pollution-heavy technology. For that reason, experts believe that China will soon become the world's number one emitter of greenhouse gases.

In addition to global warming, fossil fuels cause other environmental problems, and coal is the worst offender:

> [C]oal-burning electric power plants have fouled the air with enough heavy metals and other noxious pollutants to cause 15,000 premature deaths annually in the US alone, according to a Harvard School of Public Health study. Believe it or not, a coal-fired plant releases 100 times more radioactive material than an equivalent nuclear reactor—right into the air, too, not into some carefully guarded storage site.[16]

Other traditional fuels pose environmental problems as well. Natural gas also causes greenhouse-gas emissions. Recovering unconventional oil will pollute the water and contaminate the atmosphere. Government officials cannot agree on where to put radioactive wastes from nuclear power plants; in the meantime, utility companies store them near the plants themselves.

Summary

By relying heavily on foreign oil, America has compromised its national security. Some countries that supply oil are hostile toward us, whereas others are unstable. The money we pay for that oil enriches corrupt governments and props up regimes with bad human-rights records. Some of it ends up in the hands of insurgents and terrorists. At home, our fossil-fuel infrastructure is vulnerable to terrorist attacks. Dependence on foreign oil requires us to maintain a military presence overseas, which entangles us in local political disputes and breeds anti-American sentiment. As the supply of oil runs out, we face a growing risk of war for resources with other countries, including China and Russia. Fossil fuels emit greenhouse gases, which have been identified as a leading cause of global warming.

Alternative Energy Is Riskier Than Traditional Fuels

In 1990, General Motors Corporation introduced a concept electric car at the Los Angeles Auto Show. Seven years later, GM shipped the EV1 to test customers; Ford Motor Company bought the rights to the Th!nk electric car, developed by a Norwegian company, and began shipping versions of it in the United States; Honda shipped the EV Plus; and Toyota shipped the RAV-4 EV. For a while, the electric car was touted as the solution to U.S. dependence on foreign oil and the environmental problems created by the internal-combustion engine. By 2002, however, the electric-car boom was over. Ford stopped selling Th!nk and sold the rights to a Swiss company. GM withdrew the EV1 and began recalling the cars, all of which had been leased to customers. Honda and Toyota stopped marketing their electric cars.

Electric cars could not compete with standard-model cars because they needed to be recharged often. Critics point to these cars as an example of why alternative energy is more expensive and less reliable than the fuels we use today.

Solar and wind power cannot replace traditional fuels.

Alternative sources of energy are less attractive than traditional sources for several reasons. First of all, they are less reliable. This is an important concern because even a brief power outage can cause millions of dollars in damage. Two of the most talked-about forms of renewable energy, solar and wind power, suffer from *intermittency*, which means they cannot produce power 24 hours a day, seven days a week. H. Sterling Burnett of the National Center for Policy Analysis explains what this means for wind energy: "[W]ind farms must rely on conventional power plants to back up their supply. Wind farms generate power only when the wind is blowing within a certain range of speed. When there is too little wind, the towers don't generate power; but when the wind is too strong, they must be shut down for fear of being blown down. And even when they function properly, wind farms' average output is less than 30 percent of their theoretical capacity."[1] Wind farms also require enormous tracts of land. According to Burnett, a 1,000-megawatt wind farm, which would provide enough power for a small city, would require 192,000 acres, or 300 square miles. By comparison, a nuclear plant requires fewer than 1,700 acres, and a coal-powered plant about 1,950 acres.

Solar power also suffers from intermittency, as well as low energy density. Peter Schwartz and Spencer Ross commented: "Maybe someday we'll all live in houses with photovoltaic roof tiles, but in the real world, a run-of-the-mill 1,000-megawatt photovoltaic plant will require about 60 square miles of panes

(continues on page 74)

Conventional, Renewable, and Alternative Fuels

One of the main problems with fossil fuels is that their supply is finite. As M. King Hubbert commented:

> When we consider that it has taken 500 million years of geological history to accumulate the present supplies of fossil fuels, it should be clear that, although the same geological processes are still operative, the amount of new fossil fuels that is likely to be produced during the next few thousands of years will be inconsequential.*

For that reason, as well as the contribution that fossil fuels make to global warming, attention is turning to renewable, or "alternative," sources of energy. They include the following:

- *Biomass.* Fuel from trees, grasses, agricultural crops, or other biological material. It is carbon neutral because the carbon dioxide consumed by plants offsets the carbon dioxide emitted when it is burned. Biomass fuel is biodegradable (it breaks down into simpler compounds) and less likely to result in toxic spills or groundwater contamination.
- *Biodiesel.* Fuel made from plant oils that can be used in a diesel engine. The inventor of that engine, Rudolph Diesel, brought a car powered by peanut oil to the 1900 World's Fair in Paris. Substances such as fry oil, which is currently treated as waste material that raises disposal problems, can be used as fuel instead.
- *Geothermal energy.* The Earth's heat is a vast and untapped energy source. This energy source is reliable because heat is continuously emitted. Its price is unlikely to fluctuate because the energy supply is secured before work begins on the plant itself.
- *Hydrogen.* It is the most abundant element on Earth. Advocates believe that scientists will eventually develop fuel cells that run on hydrogen gathered from renewable sources. These fuel cells will not emit greenhouse gases.
- *Solar power.* This is a virtually limitless source of energy. Producing solar energy generates no greenhouse-gas emissions. Solar power is already widely used for small-scale applications such as road signs and pocket calculators.

- *Tidal power.* Tidal energy, like geothermal energy, is always available. Producing energy from tides generates no carbon emissions.
- *Wave energy.* Because most of the Earth's surface is covered by water, scientists are studying the physics of waves in the hopes of collecting their energy. Small-scale wave energy projects are already operational.
- *Wind energy.* Modern technology has improved on the traditional windmill. Today, wind power is a growing portion of energy production as the cost of generating this energy comes down. Recently, oil billionaire T. Boone Pickens invested part of his fortune in wind farms in Texas.

Today, we still rely heavily on "conventional," or nonrenewable, sources of energy. Several conventional fuels have been mentioned as near-term substitutes if oil and natural gas become scarce:

- *Coal.* The United States has huge reserves—by some estimates, 200 years' worth. Future technology may result in coal-based hybrid fuels or even a process by which the carbon dioxide from burning coal is sent somewhere other than the atmosphere. In the meantime, coal accounts for about one-quarter of America's energy needs, and more than half of our electric power comes from coal-burning plants.
- *Hydroelectric power.* Using water to generate power is a well-established technology, already competitive with other sources of energy.
- *Nuclear power.* Nuclear plants produce less greenhouse-gas emissions than coal- and gas-fired plants. The utility industry has produced electricity from nuclear plants for half a century. These plants have a good safety record; American plants in particular are safer in the wake of the Three Mile Island accident.

We rely on a "portfolio" of energy sources such as gasoline for cars, natural gas to heat homes, and electricity to run appliances. In the future, because of energy depletion and global warming, that portfolio is expected to include a larger proportion of alternative fuels.

* M. King Hubbert, *Nuclear Energy and the Fossil Fuels.* Presented before the Spring Meeting of the Southern District Division of Production, American Petroleum Institute, March 7–9, 1956. Houston: Shell Development Company, 1956.

(continued from page 71)
alone. In other words, the largest industrial structure ever built."[2] In addition, solar panels for buildings have a limited useful life before they must be replaced; and scientists have yet to produce a battery that would make solar energy competitive with other fuels for cars.

As the result of these practical problems, wind and solar energy have only limited potential to replace traditional sources. Paul Roberts observes:

> On average, analysts say, wind and solar renewables can provide a maximum of 20 percent of a region's power. Past that point, either the intermittency factor causes too many power disruptions, or the cost of maintaining so much backup base load becomes too high—a nonstarter for utilities trying to avoid blackouts, price increases, or anything else that might attract regulatory attention in the post-Enron era.[3]

Other alternatives have serious practical problems.

Although 100 percent renewable energy is a worthy goal, we are not even close to achieving it. Fossil fuels, usually large quantities of them, must be used in the process of generating energy from alternatives—for example, to build turbines that capture wind energy and to transport panels that collect solar energy. In fact, fossil fuels must be burned in the process of generating hydrogen power. E.R. Pat Murphy, the executive director of the Community Solution, points out that nearly all of the hydrogen that is used for power comes from "reforming," a process in which a fossil fuel undergoes a chemical reaction that releases the hydrogen it contains. The by-products of reforming, however, include carbon dioxide and other pollutants.

In addition, the leading alternative fuels require either building a new infrastructure or making significant modifications to the existing one—which took 100 years and trillions of

(continues on page 77)

QUOTABLE

"The Moral Equivalent of War": President Carter's 1977 Address on Energy Policy

Shortly after taking office, President Jimmy Carter addressed the nation to rally support for the national energy plan that he was about to propose to Congress. The plan was based on 10 fundamental principles that included economic growth, protecting the environment, fairness to all, developing new energy sources, and, most importantly, conservation. The president set out a number of goals to be reached by 1985: cutting America's oil imports in half, establishing a strategic petroleum reserve, increasing the nation's coal production by two-thirds, and using solar energy in more than 2.5 million homes.

Here are some excerpts from the president's remarks:

> Tonight I want to have an unpleasant talk with you about a problem unprecedented in our history. With the exception of preventing war, this is the greatest challenge our country will face during our lifetimes. The energy crisis has not yet overwhelmed us, but it will if we do not act quickly.
>
> We simply must balance our demand for energy with our rapidly shrinking resources. By acting now, we can control our future instead of letting the future control us....
>
> Our decision about energy will test the character of the American people and the ability of the president and the Congress to govern. This difficult effort will be the "moral equivalent of war"—except that we will be uniting our efforts to build and not destroy.
>
> I know that some of you may doubt that we face real energy shortages. The 1973 gasoline lines are gone, and our homes are warm again. But our energy problem is worse tonight than it was in 1973 or a few weeks ago in the dead of winter. It is worse because more waste has occurred, and more time has passed by without our planning for the future. And it will get worse every day until we act.
>
> The world now uses about 60 million barrels of oil a day and demand increases each year about 5 percent. This means that just to stay even we need the production of a new Texas every year, an Alaskan North Slope every nine months, or a new Saudi Arabia every three years. Obviously, this cannot continue....

(continues)

(continued)

The world has not prepared for the future. During the 1950s, people used twice as much oil as during the 1940s. During the 1960s, we used twice as much as during the 1950s. And in each of those decades, more oil was consumed than in all of mankind's previous history....

Ours is the most wasteful nation on earth. We waste more energy than we import. With about the same standard of living, we use twice as much energy per person as do other countries like Germany, Japan and Sweden....

We can't substantially increase our domestic production, so we would need to import twice as much oil as we do now. Supplies will be uncertain. The cost will keep going up....

I am sure each of you will find something you don't like about the specifics of our proposal. It will demand that we make sacrifices and changes in our lives. To some degree, the sacrifices will be painful—but so is any meaningful sacrifice. It will lead to some higher costs, and to some greater inconveniences for everyone.

But the sacrifices will be gradual, realistic, and necessary. Above all, they will be fair. No one will gain an unfair advantage through this plan. No one will be asked to bear an unfair burden....

Other generations of Americans have faced and mastered great challenges. I have faith that meeting this challenge will make our own lives even richer. If you will join me so that we can work together with patriotism and courage, we will again prove that our great nation can lead the world into an age of peace, independence and freedom.

Carter's energy policy received mixed reviews at the time, and commentators are still divided over it. Some praise him for anticipating that energy depletion would eventually cause a crisis and for understanding the economic and political importance of energy independence. Others, however, believe that he not only overstated the problem but also tried to address it with a heavy-handed regulatory approach that offered few incentives to find new sources of energy.

Source: Jimmy Carter, "The President's Proposed Energy Policy," Address to the American People, April 18, 1977. *Vital Speeches of the Day* 43, no 14, pp. 418–420 (May 1, 1977).

(continued from page 74)
dollars to build. The Government Accountability Office (GAO) estimates that it would cost $100,000 to modify a gas station to serve vehicles that run either primarily or exclusively on ethanol, which our government is encouraging the production of. The GAO also estimates that it would cost as much as $1 million to modify a gas station to accommodate cars that run on natural gas. The costs are higher still for hydrogen-powered cars. According to David Morris of the Institute for Local Self-Reliance: "To be successful, a hydrogen initiative will require the expenditure of hundreds of billions of dollars to build an entirely new energy infrastructure (pipelines, fueling stations, automobile engines). Much of this will come from public money. Little of this expenditure will directly benefit renewables."[4] Nor will a transition away from gasoline-powered cars happen overnight. The authors of the Hirsch Report found that it would take 10 to 15 years to replace half of the nation's fleet of passenger cars and 9 to 14 years to replace half the fleet of light trucks, and that replacing those vehicles would cost $2.3 trillion.

The GAO's report also raises doubts about the practicality of several other alternatives. It questions whether biodiesel will displace gasoline, stating: "The price of soybean oil is not expected to decrease significantly in the future owing to competing demands from the food industry and from soap and detergent manufacturers. . . . As a result, experts believe that the total production capacity of biodiesel is ultimately limited compared with other alternative fuels."[5] The report also notes that biomass fuel "is not commercially produced, and a number of technological and economic challenges would need to be overcome for commercial viability,"[6] and that government researchers believe we are a long way from large-scale production of liquid fuel from biomass. It concedes that coal gas-to-liquid (GTL) is technologically
(continues on page 80)

Drawbacks of Alternative Fuels

Some energy experts contend that the alternatives to oil suffer from serious—and, in some cases, insurmountable—drawbacks. Many of these alternatives require substantial consumption of fossil fuels to produce and are more expensive to produce than energy from traditional fuel sources. There are also technical and logistical problems, including these:

- *Biodiesel/biomass.* Currently, plant life is used to produce biofuels. Growing crops to produce ethanol depletes plant life, including entire forests; consumes water; and encourages continued sprawl and automobile use.
- *Geothermal energy.* There will be opposition by residents and environmental groups to the construction of plants. It is risky and expensive to look for and extract this form of energy. Once extracted, the energy is difficult to transport.
- *Hydrogen.* Hydrogen is not a fuel but a carrier of energy. Substantial amounts of energy—derived primarily from fossil fuels—must be spent to extract hydrogen from compounds, such as water, in which it is found. Despite years of work, no one has developed a workable hydrogen fuel cell for motor vehicles. Hydrogen is the lightest element, making it difficult to transport and store. It is also highly explosive; thus, the hydrogen infrastructure—which has yet to be built—will be a target for terrorists. Government and industry might invest in hydrogen power at the expense of more promising alternatives.
- *Solar power.* Because solar energy is less energy-intense than fossil fuels, it will take thousands of square miles of solar panels to meet our energy needs. There is also an intermittency problem: Solar power is unavailable at night or on cloudy days. Solar panels do not last forever and must be replaced. Solar batteries for cars are not price competitive, and little progress has been made on those batteries in the past 100 years.
- *Synthetic fuels.* "Synfuels," liquid fuels derived from coal, are little more than a tax shelter. These fuels are not "scalable," which means that they will not become cheaper as production increases.

- *Tidal power.* Only a handful of locations are capable of generating tidal power. The extreme tides that provide energy can also damage tidal power plants.
- *Wave energy.* Wave energy is capital intensive. The same rough seas that provide energy can destroy wave power plants.
- *Wind energy.* "Wind farms," despite their name, are noisy and unsightly. Wind turbines kill migrating birds, some of which are endangered species. Wind energy, like solar energy, is intermittent; as a result, traditional energy grids have to serve as a backup. Wind power is not scalable.

There are also traditional fuels that, at least in the short term, could ease the coming oil shortage. Each of them, however, presents problems:

- *Coal.* Even though coal is used to produce electric power, there is no current technology that uses it to power cars. Burning more coal has serious environmental consequences, including global warming and acid rain.
- *Hydroelectric power.* Most of the nation's rivers are already dammed up. Even if the remaining rivers are harnessed, the added power will satisfy only a small fraction of our energy needs. Hydroelectric dams are bad for the environment; they kill fish and displace people, and reservoirs eventually fill up with silt. In any event, global warming will cause water supplies to dwindle.
- *Natural gas.* Production in this country peaked in 1973, and shortages like that of the spring of 2003 are growing more serious. Much of our gas is imported, which poses the risk of supply disruptions. Natural gas is difficult to transport across oceans, and storage facilities are vulnerable to terrorist attacks. A main component of natural gas is methane, which is a potent greenhouse gas.
- *Nuclear power.* Nuclear plants are very expensive. Much of the cost of these plants, including insurance against catastrophic events, is absorbed by taxpayers rather than electricity consumers. Nuclear energy is not suitable for powering vehicles and airplanes. Nuclear plants are not pollution-free, and disposing of radioactive waste remains a problem. They are also terrorist targets. Enriched uranium, which is used in nuclear plants, can be converted into weapons.

(continued from page 77)
feasible, but finds that none of the world's production facilities are profitable and that it costs billions of dollars and takes years to build a plant.

Finally, some alternatives exist only in theory. One example is artificial nuclear fusion, which occurs naturally in stars:

> It is possible that the holy grail of limitless clean energy through the medium of nuclear power based on fusion will eventually be successful. This would sidestep the attendant problems of nuclear fission in that there would not be any shortage of the elements to fuel the process and fewer radioactive waste products. However, despite fifty years of research on this so far, the widely accepted view is that a commercially viable breakthrough is not expected for at least another forty years, if at all.[7]

Some alternatives do more harm than good.

Hydroelectric power—generating electricity from running water, which accounts for most of the world's renewable energy—has been praised as a "clean" source of energy. This form of power, however, damages the environment:

> Large dams were once welcomed by some environmentalists as a source of clean and endlessly renewable energy, but their green credentials are tarnished. The flooding that accompanies big dams in the developing world usually submerges large tropical forests. As vegetation decays, it can release lots of methane, a much more powerful greenhouse gas than carbon dioxide. Unlike giant dams in rich countries, dams in the tropics must endure the ravages of monsoons. One common result is silting up, which may cut the original generating capacity by 70 percent or even 80 percent within a few decades.[8]

Dams also make rivers less habitable for fish and other creatures, and reservoirs created by dams become breeding grounds for mosquitoes that spread diseases such as malaria and the West Nile virus.

Although unconventional oil is abundant, extracting it has serious environmental consequences. The GAO says about fuel from oil sands: "The production process uses large amounts of natural gas, which generates greenhouse gases when burned. In addition, large-scale production of oil sands requires significant quantities of water, typically producing large quantities of contaminated wastewater, and alter the natural landscape."[9] It says that heavy oil production "presents environmental challenges in its consumption of other energy sources, which contributes to greenhouse gases, and potential groundwater contamination from the injectants needed to thin the oil enough so that oil will flow through pipes."[10] It also says that producing fuel from oil shale "is energy-intensive, requiring other energy sources to heat the shale to about 900 to 1,000 degrees Fahrenheit to extract the oil. Furthermore, oil shale production is projected to contaminate local surface water with salts and toxins that leach from spent shale."[11]

Critics call ethanol the worst alternative of all. Ethanol is a fuel for vehicles that is derived from corn, which grows in abundance in the Midwest. The federal government encourages farmers to grow corn by offering tax breaks. The ethanol program, however, has diverted corn crops from food to fuel production. In 2007, Lester Brown of the Earth Policy Institute found that corn prices had doubled in the previous year, wheat futures were trading at their highest level in 10 years, and rice prices were on their way up as well. Brown warned of more price increases to come:

> Some 16 percent of the 2006 US grain harvest was used to produce ethanol. With 80 or so ethanol distilleries now under

> construction, enough to more than double existing ethanol production capacity, nearly a third of the 2008 grain harvest will be going to ethanol. Since the United States is the leading exporter of grain, shipping more than Canada, Australia, and Argentina combined, what happens to the US grain crop affects the entire world. With the massive diversion of grain to produce fuel for cars, exports will drop.[12]

Ethanol production also raises ethical questions. George Monbiot, a columnist for the British newspaper the *Guardian*, says: "The market responds to money, not need. People who own cars have more money than people at risk of starvation. In a contest between their demand for fuel and poor people's demand for food, the car-owners win every time."[13]

Alternatives will lower our standard of living.

Activists are calling for immediate action to shift the nation from traditional to alternative fuels. A "crash program," however, would likely drive up the price of energy. Hasty government action could lock us into inefficient forms of energy and close the door to better alternatives that might be found in the future. In addition, focusing too much on conservation could make our economy less vibrant. Peter Huber and Mark Mills observe: "Slower trips, dimmer bulbs, smaller refrigerators, and such aren't more efficient; they're slower, smaller, darker—they nudge us toward a less frenetic, peripatetic, and physically expansive way of life. Perhaps this is a good thing. But it is not more efficient, it is more sedentary, calm, and quiet—in short, more lethargic."[14]

Although much has been said about the negative consequences of fossil-fuel use, those fuels make our standard of living possible. Bjorn Lomborg, a professor at the Copenhagen Business School, says: "Cheaper fuels would have avoided the deaths of a significant number of the 150,000 people who have died in the United Kingdom since 2000 due to cold winters. Fossil fuels allow food to be grown cheaply, and we can have access to fruits and vegetables year-round, which probably has reduced cancer rates by at least 25 percent."[15] Lomborg also reminds us of the harshness of life without these fuels: "Two and a half billion people use biomass such as wood, waste, and dung to cook and keep warm. For many Indian women, searching for wood costs three hours each day, as they sometimes walk more than six miles per day. It also causes excessive deforestation. About 1.3 million people—mostly women and children—die each year due to heavy indoor-air pollution."[16]

In addition, many activists are biased against nuclear energy, even though it represents an excellent alternative to fossil fuels. Supporters of nuclear power point out that half of America's electricity is generated by coal-fired power plants, which emit greenhouse gases and discharge other pollutants that cause acid rain and a variety of health problems. Michael Totty, a reporter at the *Wall Street Journal*, observes: "Nuclear power plants, on the other hand, emit virtually no carbon dioxide—and no sulfur or mercury either. Even when taking into account 'full life-cycle emissions'—including mining of uranium, shipping fuel, constructing plants and managing waste—nuclear's carbon-dioxide discharges are comparable to the full life-cycle emissions of wind and hydropower and less than solar power."[17] Nuclear power also has an undeserved reputation for being unsafe because of media coverage of the leak at Three Mile Island and films such as *The China Syndrome*, which depicted a near-disaster at a fictional plant. The National Energy Policy Development Group responded to concerns about safety: "Since the Three Mile Island incident in 1979, the nuclear industry's safety record has significantly improved. This safety record has been achieved through a defense-in-depth philosophy accomplished by way of engineering design, quality construction, safe operation, and emergency planning."[18]

Some people believe that environmental activists' opposition to nuclear energy reflects general hostility to the way we live. They accuse activists of using energy depletion to force a left-wing economic and social agenda. In a speech on the Senate floor, Senator James Inhofe of Oklahoma said: "For them, a 'pro-environment' philosophy can only mean top-down, command-and-control rules dictated by bureaucrats. Science is irrelevant—instead, for extremists, politics and power are the motivating forces for making public policy."[19] Furthermore, many activists are wealthy enough to afford higher taxes and energy prices that result from restrictions on oil drilling and the construction of nuclear plants. Inhofe reminds us that working-class and poor Americans are not as fortunate: "Energy suppression, as official government and nonpartisan private analyses have amply confirmed, means higher prices for food, medical care, and electricity, as well as massive job losses and drastic reductions in gross domestic product, all the while providing virtually no environmental benefit. In other words: a raw deal for the American people and a crisis for the poor."[20]

Summary

Even though the alternative energy industry has existed for decades, it has yet to become a commercial success. Despite billions of dollars in government aid, alternative fuels are still more expensive than traditional fuels and account for a small share of overall energy consumption. Alternative fuels will not end our dependence on fossil fuels, because current technology still requires substantial amounts of fossil fuels to produce alternatives. In addition, many alternatives raise serious environmental concerns. Some, especially ethanol fuel, create even worse problems than those they were designed to solve. Others, such as hydrogen-powered cars, may never become a reality. Traditional fuels—including nuclear power, which is safe and reliable—allow humans with access to them to live in comfort.

Government Must Address the Energy Crisis

For the past 30 years, our government has relied on market forces—supply and demand—to encourage conservation and find substitutes for fossil fuels. When oil production peaks, however, market forces might act too slowly or in too uncoordinated a fashion to avert serious disruption. Up to now, a "hands-off" approach by government has made sense: Every time the price of oil rose sharply, it eventually went back down. The authors of the Hirsch Report, however, warn: "The peaking of world oil production could create enormous economic disruption, as only glimpsed during the 1973 oil embargo and the 1979 Iranian oil cut-off."[1] They also say that, although government cannot eliminate the adverse effects of sudden and severe oil disruptions, it can minimize those effects.

The energy market is not "free."

No market is perfect, and that is especially true in the case of energy. To begin with, markets put too low a price on fossil fuels. Paul Roberts explains: "You and I may know that hydrocarbons cost us dearly, in terms of smog, climate change, corruption, and instability, not to mention the billions spent defending the Middle East. But because these 'external' costs aren't included in the price of a gallon of gasoline, the market sees no reason to

U.S. House Resolution on Peak Oil

In 2005, Roscoe Bartlett, a Republican U.S. representative from Maryland, founded the House Peak Oil Caucus. For a number of years, Bartlett has warned of the consequences of peak oil, sometimes from the House floor.

During the 109th Congress, Bartlett and 14 of his colleagues introduced a nonbinding resolution (H.R. 507) that would express "the sense of the House of Representatives" regarding peak oil. It read as follows:

> Whereas the United States has only 2 percent of the world's oil reserves;
>
> Whereas the United States produces 8 percent of the world's oil and consumes 25 percent of the world's oil, of which nearly 60 percent is imported from foreign countries;
>
> Whereas developing countries around the world are increasing their demand for oil consumption at rapid rates; for example, the average consumption increase, by percentage, from 2003 to 2004 for the countries of Belarus, Kuwait, China, and Singapore was 15.9 percent;
>
> Whereas the United States consumed more than 937,000,000 tonnes of oil in 2004, and that figure could rise in 2005 given previous projection trends;
>
> Whereas, as fossil energy resources become depleted, new, highly efficient technologies will be required in order to sustainably tap replenishable resources;
>
> Whereas the Shell Oil scientist M. King Hubbert accurately predicted that United States domestic production would peak in 1970, and a growing number of petroleum experts believe that the peak in the world's oil

find something other than oil, gas, or coal."[2] As a result, alternatives seem more expensive than they are, which makes them less competitive with traditional fuels. Nor does the price of energy reflect environmental effects such as air and water quality, health problems resulting from pollution, or the extinction of plant and animal species. The most serious environmental effect is the discharge of huge amounts of carbon dioxide and other greenhouse gases into the atmosphere.

production (Peak Oil) is likely to occur in the next decade while demand continues to rise;

Whereas North American natural gas production has also peaked;

Whereas the United States is now the world's largest importer of both petroleum and natural gas;

Whereas the population of the United States is increasing by nearly 30,000,000 persons every decade;

Whereas the energy density in one barrel of oil is the equivalent of eight people working full time for one year;

Whereas affordable supplies of petroleum and natural gas are critical to national security and energy prosperity; and

Whereas the United States has approximately 250 years of coal at current consumption rates, but if that consumption rate is increased by 2 percent per year, coal reserves are reduced to 75 years: Now, therefore, be it

Resolved, That it is the sense of the House of Representatives that—

(1) in order to keep energy costs affordable, curb our environmental impact, and safeguard economic prosperity, including our trade deficit, the United States must move rapidly to increase the productivity with which it uses fossil fuel, and to accelerate the transition to renewable fuels and a sustainable, clean energy economy; and

(2) the United States, in collaboration with other international allies, should establish an energy project with the magnitude, creativity, and sense of urgency of the "Man on the Moon" project to develop a comprehensive plan to address the challenges presented by Peak Oil.

The resolution died in the House Energy and Commerce Committee.

Not only do markets underprice energy, but consumers are also unaware of the cost savings resulting from conservation measures. A consumer who buys a car or an appliance usually considers its purchase price but not the cost of operating it for its useful life. Often, however, the initial cost saving is offset by the higher cost of energy. Paul Roberts believes that government can make citizens more "energy literate." He says: "Governments would need to rewrite everything from building codes to tax laws, in order to encourage investment in efficiency upgrades. Industries would need to rethink the way they do their energy accounting and, in particular, incorporate life-cycle energy costs into the bidding process for capital projects. But the payoff would be enormous."[3]

In any event, there is no such thing as a free market in energy. Ross Gelbspan explains:

> Energy is probably the most highly regulated commodity there is—from price fixing by OPEC, to utility protection regulations, to the regulating of automobile mileage standards, to government subsidies for coal and oil and a host of other local, state, and federal regulations. Beyond that, the regulations that inform the production and distribution of energy are rigged to conform to the requirements of the coal, oil, and auto industries as they are currently configured. They provide little space for efficiency innovations and even less for renewable energy technologies.[4]

The energy market has become even *less* free because state-owned companies now dominate the oil industry and make decisions, such as how much to produce and whom to sell it to, on the basis of their national interest rather than making a profit.

America has no comprehensive energy policy.

Critics describe America's current energy policy as a collection of ad hoc measures, many of which are aimed at benefiting local economic interests. They believe that the time has come for our government to adopt a comprehensive energy policy. There are a number of reasons why the government should step in. To begin with, if government fails to protect the public interest, no one else will. Left unregulated, the energy companies will continue to make decisions that emphasize immediate profits while ignoring long-term problems such as energy depletion and global warming.

Government should also encourage conservation for the long term, even when prices are low and saving energy is not perceived as a high priority. Roberts observes:

> Paradoxically, conservation's great success was also its downfall. As oil prices fell to ten dollars a barrel, few Western consumers saw any reason to continue conserving. In Europe and Japan, where energy security remained a critical issue, government kept fuel taxes high, to discourage oil imports—a policy that has, by and large, worked. But in the United States, where raising taxes is anathema (and where the politically connected domestic oil industry was desperate to see demand increase), political leaders declared the energy crisis over.[5]

European countries use half as much energy per capita and half as much per unit of economic activity as the United States does, in large part because their governments have imposed high taxes on energy for many years.

As oil production inevitably reaches a peak, only the government can manage an economy-wide transition to alternatives. Some believe that this effort will be an undertaking on the scale of America's mobilization during World War II. Planning for peak oil also involves a time frame longer than what corporate executives are accustomed to. The authors of the Hirsch Report recommend that government take steps to deal with peak oil long before it actually occurs. They also say: "Intervention by governments will be required, because the

Portland, Oregon, Prepares for Peak Oil

According to the London-based Oil Depletion Analysis Centre, the most advanced U.S. city preparing for peak oil is Portland, Oregon. The Portland Peak Oil Task Force, a group of 12 citizens with varied backgrounds, published a report* that was approved by City Council in 2007. Key priorities identified in that report include:

- Reducing total oil and natural gas consumption by 50 percent over the next 25 years.
- Informing citizens about peak oil and developing community-based solutions.
- Encouraging business, government, and community leaders to plan for peak oil and develop policies for dealing with it.
- Supporting land-use patterns that will reduce the need for transportation, promote walkability, and provide easy access to services.
- Designing infrastructure that will support alternatives to driving and promote the efficient movement of freight, and prevent investments in infrastructure that will not be prudent in light of the coming fuel shortages and higher prices.
- Encouraging energy-efficient and renewable transportation choices.
- Expanding energy-efficiency programs for buildings, and providing incentives that would improve the efficiency of new and existing buildings.
- Preserving farmland, and expanding local production of food.
- Identifying and promoting sustainable business opportunities.
- Redesigning the social "safety net" to protect vulnerable and marginalized citizens.
- Preparing emergency plans for sudden and severe energy shortages.

* Portland Peak Oil Task Force. *Descending the Oil Peak: Navigating the Transition from Oil and Natural Gas*. Portland, OR, 2007. http://www.portlandonline.com/shared/cfm/image.cfm?id=145732.

economic and social implications of oil peaking would otherwise be chaotic."[6]

Finally, government should take the lead in cutting fossil-fuel consumption, for example, by making public buildings energy efficient and buying alternative-fuel vehicles for government fleets. Public officials themselves should set an example, as President Jimmy Carter did when he had solar panels installed on the White House. Many also believe that America, the world's top energy-consuming nation, should become a world leader in energy conservation as well. Roberts explains: "The United States is the only country with the economic muscle, the technological expertise, and the international standing truly to mold the next energy system. If the U.S. government and its citizens decided to launch a new energy system and have it in place within twenty years, not only would the energy system be built, but the rest of the world would be forced to follow along."[7]

Our government encourages fossil-fuel consumption.

Whereas the market may underprice oil, our government's policies make it cheaper still by subsidizing the energy industry. Oil and gas companies receive a variety of tax breaks. According to the advocacy group Public Citizen, the Energy Policy Act of 2005 alone contained $6 billion in tax breaks and subsidies for oil and gas companies, $9 billion for the coal industry, and $12 billion for producers of nuclear power. Public Citizen called these subsidies a payback to energy companies that have contributed $115 million to candidates for federal office since 2001. Three-quarters of that money went to Republicans, who controlled the White House and both houses of Congress when the legislation passed. Energy companies were given private access to lawmakers, as well as members of National Energy Policy Development Group, whose recommendations went into the 2005 legislation.

Even after three energy crises and despite worldwide concern over greenhouse-gas emissions, our government continues to encourage Americans to drive cars:

> Taxes on the purchase price of vehicles are mostly set at the sales tax rates of different states, varying from zero to about 8 percent. But the bulk of the revenue is earmarked for the construction of highways and parking lots—less than 20 percent goes to support mass transit. Indeed, about 60 percent of funding for highways comes directly from fuel and vehicle taxes, thus maintaining the system of automobility as a well-oiled machine.[8]

The damaging effects of this national policy include suburban sprawl, deaths in traffic crashes, and childhood obesity, along with excessive consumption of fossil fuels.

Alternative energy is a wise investment.

Policies that shift our economy to alternative fuels will yield significant benefits in the long run. To begin with, those fuels are renewable, which lessens the chances that the supply will run out. Because alternatives are available in this country, foreign governments cannot manipulate their price or cut off our supply. Alternative energy can create jobs. In 2000, renewable-energy analyst Scott Sklar estimated that $1 million spent on oil and gas exploration created only 1.5 jobs and $1 million on coal mining created 4.4 jobs. By contrast, $1 million spent on making solar water heaters created 14 jobs, $1 million spent on manufacturing solar panels created 17 jobs, and $1 million spent on producing electricity from biomass created 23 jobs. These are good-paying jobs, such as welding, driving trucks, and operating construction equipment, and they cannot to be outsourced to other countries.

A switch to renewables could reduce the cost of all forms of energy. According to the Union of Concerned Scientists: "Even when they cost more, renewable technologies help stabilize electricity prices: because their operating costs are low and their fuel is free, they create competitive pressure, which acts to restrain the price of fossil fuels like natural gas."[9] Another benefit of alternative energy is that it is compatible with a "distributed" or decentralized energy infrastructure, which is less vulnerable to disruption and would waste less power in the process of generating and delivering it.

Finally, in recent years, the cost of traditional fuels has risen to a level that makes alternative fuels competitive. Former vice president Al Gore elaborates:

> When I first went to Congress 32 years ago, I listened to experts testify that if oil ever got to $35 a barrel, then renewable sources of energy would become competitive. . . . And sure enough, billions of dollars of new investment are flowing into the development of concentrated solar thermal, photovoltaics, windmills, geothermal plants, and a variety of ingenious new ways to improve our efficiency and conserve presently wasted energy.
>
> And as the demand for renewable energy grows, the costs will continue to fall. Let me give you one revealing example: the price of the specialized silicon used to make solar cells was recently as high as $300 per kilogram. But the newest contracts have prices as low as $50 a kilogram.[10]

Gore also compared solar cells to computer chips, which are also made out of silicon: The price for the same performance from chips has fallen by 50 percent every 18 months for a 40-year period.

In any event, switching to alternative fuels will not cause economic disruption. In 1997, a group of more than 2,500 economists, including eight Nobel Prize winners, concluded that the American economy could weather the change to alternative fuels and even improve productivity in the long run. Vijay

Vaitheeswaran disputes claims that we cannot afford to build a renewable energy infrastructure. In particular, he takes issue with the often-cited estimates of $100 billion or more for phasing in a hydrogen infrastructure: "They assume that today's gasoline infrastructure must be duplicated from day one, which is not the case. Experience with the introduction of diesel in America and unleaded gasoline in Germany shows that if even if only

THE LETTER OF THE LAW

The Energy Independence and Security Act of 2007

The Energy Independence and Security Act of 2007 (Public Law 110–140) was signed into law by President George W. Bush on December 19, 2007. The highlight of the legislation is Title I, which tightens CAFE standards. The so-called "Ten-in-Ten Fuel Economy Act" sets a minimum standard of 27.5 mpg for the total fleet of passenger and non-passenger automobiles for the 2011 model year. That standard rises to at least 35 mpg for the 2020 model year. Title I authorizes the Transportation Department to create a CAFE credit-trading program, under which automakers that exceed fuel economy standards earn credits that they can sell to automakers that fall below the standards. It also directs the Transportation Department to develop a system that rates new cars on the basis of fuel economy and greenhouse-gas emissions, and provides funding for alternative technology such as hybrid and electric vehicles.

Title II directs the Environmental Protection Agency to adopt rules to ensure that fuel used in cars and trucks contains renewable or alternative fuels. The goal is a 20 percent reduction in greenhouse-gas emissions. It also provides incentives for ethanol, biodiesel, and biogas, as well as developing an infrastructure that would support these alternatives.

Titles III through V aim to achieve greater energy efficiency in lighting and appliances and in residential, commercial, and government buildings. It imposes various requirements that will force the government to lead the way toward energy efficiency, such as buying low-emissions vehicles and using alternative energy inside government buildings.

Title VI provides incentives for business and higher education to develop alternative energy technology.

15 percent of gas stations offer it, a new fuel can become widely accepted."[11]

Government programs can ease energy crises.

After the Arab oil embargo, Congress passed the Energy Policy and Conservation Act of 1975,[12] which required new cars to average at least 18 mpg. The so-called CAFE standards forced

Title VII directs the Department of Energy to conduct research and development into carbon capture and sequestration—storing carbon dioxide underground rather than allowing it to escape into the atmosphere.

Title VIII directs the Department of Energy to conduct a media campaign to persuade Americans to use energy more efficiently and to promote the national-security benefits of energy efficiency. Section 806, which is nonbinding, expresses the sense of Congress that, by 2025, one-quarter of the energy used in the United States should come from renewable resources.

Title IX directs various federal agencies to work with foreign countries to develop clean and efficient energy technology and to make energy independence a higher foreign-policy priority.

Title X directs the Department of Labor to establish a "green jobs" training program.

Title XI creates an Office of Climate Change and Environment within the Department of Transportation.

Title XII provides for loans to small businesses to buy equipment that would make their operations more energy efficient, and for venture capital for small businesses in the renewable energy industry.

Title XIII declares a policy of modernizing the nation's electric-energy infrastructure, including the adoption of "smart grid" technology. The smart grid is an advanced version of the electric meter that identifies consumption in more detail and communicates that information back to the utility for monitoring and billing purposes. The goal is to encourage customers to shift energy use to off-peak hours when demand is lower, thus saving money for the customer and the utility company.

automakers to make more fuel-efficient cars years earlier than they otherwise would have. Joan Claybrook, the head of Public Citizen, told a U.S. Senate committee that, unless the government stepped in, there was no guarantee that automakers would have implemented fuel-efficiency technology they had already developed. Claybrook called CAFE "a smashing success," pointing out that it made American cars 82 percent more fuel-efficient between 1978 and 1985. She also said that CAFE was a major factor in breaking OPEC's stranglehold on oil prices and cutting rampant inflation during the early 1980s.

CAFE was not the only successful response to the "oil shocks" of the 1970s. The authors of the Hirsch Report cited the consensus of economists that "appropriate actions" taken by the government included the 55 mph speed limit, reorganization of the federal energy bureaucracy, greatly increased research and development related to energy, creation of the Strategic Petroleum Reserve, stronger energy-efficiency standards and building codes, and the establishment of the International Energy Agency.

In an effort to encourage production, Germany passed laws that require utility companies to buy electricity generated from renewable sources—including from individuals who install solar panels or wind turbines. Author Craig Morris argues that the legislation not only "leveled the playing field" for small producers but has encouraged greater production along with reliability: "Retail rates have remained stable, not skyrocketed, and there have been no rolling brownouts. On the contrary, Germany has only around 20–30 minutes of power outages on average each year—among the lowest blackout figures in the world. Renewables now make up 13 percent of the country's electricity supply—and this share is rising by around two percent per annum."[13] The approach of purchasing energy from individuals with solar panels or wind turbines is called "net metering" because the customer is billed for the net amount of power consumed—that is, the amount the utility company sends the customer minus what the customer sends back to the utility company.

Those who are pushing for relief from polluting fossil-fuel energy often suggest converting all or part of the nation's energy industry to alternative energy sources such as solar or wind power, both of which are renewable and do not pollute. Opponents of this proposal, however, note that the technology for both still needs to overcome intermittency—meaning that they only work when, respectively, the sun is shining or the wind is blowing.

The German government also offers subsidies and loans to companies that generate power from renewable sources. Some believe that government aid can increase the chances that alternative energy will have a large enough market to make it competitive. Roberts believes this could happen with wind power:

> As wind takes a greater share of the global power market, an interesting dynamic kicks in. Growing numbers drive down costs, and growth itself becomes a target: a power company's profits come to depend more and more on how fast it can

> expand its wind portfolio by adding new machines and new farms. In this way, wind becomes yet another battleground where ruthless energy giants vie for market share, cut costs, and push wind power into new markets—just the impetus that this new power source needs.[14]

Many believe that government aid will increase the market for solar energy and make it competitive as well.

Summary

Market forces cannot solve our energy problems, because the energy market is heavily influenced by government policies, here as well as abroad. The price of fossil fuels does not reflect their true costs, such as the environmental damage they cause. Their price is further distorted by large subsidies paid to companies that produce them. Our government lacks a comprehensive energy policy, which we need to cope with the eventual peaking of oil production and to combat global warming. A national energy policy must encourage conservation as well as efforts to find and develop alternatives to fossil fuels. Government also might provide the impetus to make alternative fuels competitive. Some programs have succeeded in cutting fossil-fuel use, and the government can, and should, do more.

A Free Market Is the Soundest Energy Policy

In 1776, the economist Adam Smith wrote *The Wealth of Nations*, in which he argued that free markets act as an "invisible hand": When people act in their economic self-interest, their decisions, taken together, promote the common good. Advocates believe that Smith's "invisible hand" theory is just as valid today in our "laissez-faire" system—a system that desires minimal state intervention in economic matters. Authors Peter Huber and Mark Mills remark: "The case for laissez faire today is not that the market knew best when it gave us the internal combustion engine, though it did—it is that the market now knows how to change that engine beyond recognition, and it will."[1] Huber and Mills go on to say: "The best thing U.S. policy makers can do is step out of the way and let the market find its own way to the extraordinary future that now beckons."[2]

Government cannot solve energy crises.

History is filled with examples of governments that try to manage the economy. Most of these "top-down" approaches, such as the Soviet Union's "five-year plans," are now considered failures. That is especially true of energy policy. Vijay Vaitheeswaran observes that, until recently, governments the world over had considered energy too "strategic" to be left to market forces. He goes on to say, "In many ways, they have ensured that oil, gas, and electricity operated outside proper market principles. Decades of mismanagement, inefficiency, unnecessary pollution, and excessively high costs have been the result."[3] In this country, utility companies were "regulated monopolies" for much of the twentieth century. Consumers had no choice who to buy power from, so government regulators protected them by telling utility companies how much they could charge. At the same time, regulators set rates high enough to allow utility companies to make the investments they saw fit and still earn a profit. This approach, however, encouraged utilities to invest in expensive power plants that rely on obsolete technology and large power grids that are vulnerable to disruptions.

Critics argue that government regulators lack the expertise to make important decisions in specialized areas such as energy exploration; also, because they are public employees, they are not accountable to shareholders if their decisions are wrong. Huber and Mills cite one example of bad decision-making. During the 1990s, energy activists persuaded regulators to require higher-efficiency ballasts for fluorescent lights. The regulators, however, misunderstood the economics of energy efficiency and were unaware that the state of the art in lighting was advancing. Huber and Mills comment: "In other words, those who invested heavily in the 'efficient' technology of the 1990s wasted their money. They would have done far better—and achieved far more efficiency—by just stashing away the cash for a few years until the next great thing in efficiency materialized."[4]

Jerry Taylor, the director of natural resource studies at the Cato Institute, says that government regulators are beginning to repeat the mistakes of the past by ordering utilities to generate power from alternative sources. He argues that if alternatives were competitive with gas and coal, utilities would use them without waiting for the government to tell them to. Taylor goes on to say:

> What about this exponential increase in renewable energy, particularly wind power, we keep hearing about? Well, it doesn't take much to show huge increases in market share when current production is so infinitesimal. But the main reason for the growth in renewables isn't improving economics, it's increasingly bossy politicians. Of the 5,356 megawatts of renewable energy production currently on the drawing board, only 291 megawatts would be generated voluntarily. The remainder is being built because state legislators have ordered it to be built.[5]

Finally, some believe that the government often can achieve better results by educating the public than through heavy-handed regulation. One example is the ENERGY STAR program, a voluntary labeling program begun in 1992 that is overseen by the Environmental Protection Agency and the Department of Energy:

> Through its partnerships with more than 12,000 private and public sector organizations, ENERGY STAR delivers the technical information and tools that organizations and consumers need to choose energy-efficient solutions and best management practices. ENERGY STAR has successfully delivered energy and cost savings across the country, saving businesses, organizations, and consumers about $16 billion in 2007 alone. Over the past decade, ENERGY STAR has been a driving force behind the more widespread use of such

(continues on page 104)

Recommendations of the National Energy Policy Development Group

Shortly after taking office in 2001, President George W. Bush convened the National Energy Policy Development Group.

In its final report,* the panel recommended that the executive branch of government take a number of steps to make energy more available and the supply more reliable. Key recommendations included:

- Reducing the impact of high energy costs on low-income Americans; for example, by increasing funding for the Low Income Home Energy Assistance Program and the Weatherization Assistance Program.
- Directing federal agencies to act faster on issuing permits and taking other actions needed to approve energy-related projects; and reducing the time and cost involved in licensing hydroelectric power plants.
- Expanding the ENERGY STAR program beyond office buildings to include schools, retail buildings, health care facilities, and homes; and extending the ENERGY STAR labeling program to include additional products and services.
- Recommending new CAFE standards and proposing strategies to reduce congestion.
- Working with Congress to create a tax credit for fuel-efficient vehicles, including a temporary income tax credit for buyers of hybrid fuel cell vehicles.
- Taking steps to expand oil drilling in this country, including economic incentives for offshore oil and gas development, continued approval of oil and gas exploration on the Outer Continental Shelf, and possible additional oil and gas development in Alaska's National Petroleum Reserve. In addition, the secretary should work with Congress to authorize exploration and development in the Arctic National Wildlife Refuge in Alaska.
- Proposing comprehensive electricity legislation that amends or abolishes regulations that inhibit competition and make the supply of power less reliable.

- Continuing to support advanced clean-coal technology by investing $2 billion over 10 years to fund research, making the existing research-and-development tax credit permanent, and exploring regulatory approaches that will encourage new technology.
- Supporting the expansion of nuclear energy as a major component of our national energy policy.
- Extending and expanding tax credits for electricity produced using wind and biomass, and continuing the fuel tax exemption for ethanol fuel.
- Developing next-generation energy, including hydrogen and fusion, and supporting legislation reauthorizing the Hydrogen Energy Act.
- Removing constraints on the interstate electricity transmission grid, establishing a national grid, and identifying transmission bottlenecks and measures to remove them.
- Working closely with Canada and Alaska to expedite the construction of a pipeline to deliver natural gas to the lower 48 states.
- Taking steps to ensure that America has adequate refining capacity to meet consumers' needs, including giving refinery owners clearer directions as to what the government requires and streamlining the process of getting a permit.
- Making energy security a priority of America's trade and foreign policy.
- Reaffirming that the Strategic Petroleum Reserve is designed for addressing an imminent or actual disruption in oil supplies, and not for managing prices.

Even though President Bush left office in January 2009, many of his energy panel's recommendations will be debated by Congress and the Obama administration in the years to come.

* National Energy Policy Development Group. *National Energy Policy: Reliable, Affordable, and Environment Sound Energy for America's Future*. Washington, D.C., 2001.

(continued from page 101)

> technological innovations as efficient fluorescent lighting, power management systems for office equipment, and low standby energy use.[6]

Energy regulation failed in the past.

After the Arab oil embargo, Presidents Richard Nixon, Gerald Ford, and Jimmy Carter relied heavily on regulation to hold down prices and encourage conservation. Experts now conclude that their approach likely made the problem worse. According to the authors of the Hirsch Report: "Economists have debated whether the economic problems of the 1970s were due to the oil supply disruptions or to inappropriate fiscal, monetary, and energy policies implemented to deal with them. The consensus is that the disruptions would have caused economic problems irrespective of fiscal, monetary, and energy policies, but that price and allocation controls exacerbated the impacts in the U.S. during the 1970s."[7] The list of policies that the economists judged "inadvisable" includes price and allocation controls, excessive regulation, de facto gasoline rationing, and excess-profits taxes and similar policies aimed at "greedy energy companies."

The California energy crisis of 2000–2001 is cited as a classic example of how regulation makes energy crises worse. During the 1970s, state authorities, believing that demand for electricity would remain flat, halted the construction of new power plants. In 1996, the state ordered utility companies to sell their power plants to independent companies and buy all of their power at the going price on the spot market. When supply failed to keep up with demand, energy traders such as Enron Corporation took advantage of the situation and charged utility companies steep prices, which, under California law, could not be passed along to consumers. The energy crisis bankrupted two large utility companies and forced the state to bail out a third company. The National Energy Policy Development Group offered its assessment:

> The California electricity crisis is not a test of the merits of competition in electricity markets. Instead, it demonstrates that a poorly designed state retail competition plan can have disastrous results if electricity supply does not keep pace with increased demand. At heart, the California electricity crisis is a supply crisis. California allowed demand to outstrip supply, and did little to lower barriers to entry through reform of an inflexible siting process.[8]

Environmental laws, some of which were passed before the oil shocks of the 1970s, have aggravated the nation's energy problems. According to the Energy Information Administration, more than three-quarters of this country's energy resources are under federal land. Congress, however, has made large swaths of that land off-limits to oil and gas exploration. Environmental restrictions have also made it difficult to build nuclear power plants. Peter Schwartz and Spencer Reiss, who favor nuclear power, recently wrote:

> Believe it or not, Three Mile Island wasn't the ultimate nightmare; that would be Shoreham, the Long Island power plant shuttered in 1994 after a nine-year legal battle, without ever having sold a single electron. Construction was already complete when opponents challenged the plant's application for an operating license. Wall Street won't invest billions in new plants ($5.5 billion in Shoreham's case) without a clear path through the maze of judges and regulators.[9]

In addition to environmental laws, the widespread "not in my backyard" mentality hampers the construction of new power plants. Columnist Thomas Sowell observes that such opposition has a downside:

> Somehow politicians must make it seem possible to get benefits without paying costs. But if we are too squeamish to build a

> dam and inconvenience some fish or reptiles, too aesthetically delicate to permit drilling for oil out in the boondocks and too paranoid to allow nuclear power plants to be built, then we should not be surprised if there is not enough electricity to supply our homes and support a growing economy.[10]

Deregulation increases energy supply.

For the past several decades, lawmakers have eliminated many regulations that affect markets. Vaitheeswaran says deregulation has been a success: "As the experience of the past two decades in telecommunications and computing has shown, the most powerful effect of deregulating an industry can be to open the door to venture capital, nimble entrepreneurship, and technological innovation that allows the previously unimaginable to happen."[11] During the 1980s, Prime Minister Margaret Thatcher's government deregulated Britain's energy industry. The British model has become a role model for market reformers everywhere. During the 10 years after that country deregulated its electricity market, wholesale power costs fell by one-third and retail electricity rates also dropped significantly, without any loss of reliability. In the United States, the federal government and a number of states have moved to deregulate the electric-power industry. Advocates consider Pennsylvania and Texas to be among the leaders.

Mainstream economic theory holds that when something becomes scarce, its price increases. Higher prices, in turn, cause producers to make more and consumers to either use less or look for substitutes. Energy is no exception. The price of oil sends a signal to producers. High prices encourage energy companies to look for new sources and to develop alternatives. Recently, T. Boone Pickens announced his intention to build thousands of wind turbines in the Texas Panhandle. These are the first step toward building wind farms across the Great Plains, as well as solar-energy collectors in the desert Southwest. Pickens's proposed wind farm is four times the size of the world's largest such operation and will eventually supply enough electricity to power a million homes. "Don't get the idea that I've turned green,"

Pickens told a reporter from the British newspaper the *Guardian*. "My business is making money, and I think this is going to make a lot of money."[12]

At the same time, businesses have designed more energy-efficient power plants, homes, vehicles, and appliances. The result has been ever-increasing energy efficiency—1 percent per year, by some estimates. Paul Roberts elaborates: "Today's appliances generate more comfort, entertainment, and other services than they did in 1970, for about half of the energy costs. Between 1975 and 2000, even as the American economy grew by nearly 50 percent, our 'energy intensity'—the amount of energy needed to produce a dollar of GDP—fell by 40 percent, largely through improved technology, policies, and marketing methods."[13] Even greater efficiency can be achieved in the future. Roberts points out, for example, that less than one-quarter of energy used in a standard stove actually reaches the food.

For these and other reasons, many people believe that market forces, not drastic conservation measures, represent the best hope for solving the energy crisis. Vaitheeswaran argues:

> And yet there are now more proven resources of petroleum than three decades ago; more food is produced on earth than ever before; and the last decade has seen the greatest economic expansion in history. How was this miracle possible? The answer lies in understanding that the stock of available resources is not constant. Fears of oil scarcity led to investment that discovered better and cheaper ways to produce more oil; it also led to inventions like more efficient engines that burned less oil, and to the substitution of some petroleum-based chemicals with newly developed alternatives.[14]

The government wastes tax dollars on alternatives.

Since the oil crises of the 1970s, the government has taken a variety of steps aimed at encouraging the production of alternative fuels. Critics call them a waste of tax dollars. In 2002, researchers at the Cato Institute reported that, since the Department of Energy was created, the federal government has spent more than

Should Drilling in Alaska Be Expanded?

For years, there has been intense debate over whether Congress should allow drilling for oil in environmentally sensitive areas of Alaska. In 1980, the Alaska National Interest Lands Conservation Act expanded the Arctic National Wildlife Refuge (ANWR) from 9 million acres to 19 million acres and designated 8 million acres as wilderness. Congress left open the question of drilling for oil in a 1.5-million-acre Arctic coastal plain area of ANWR, which probably contains significant oil and gas resources.

Section 1002 of the 1980 legislation directed the Department of the Interior to conduct geological and biological studies of the Arctic coastal plain, which is referred to as "the 1002 Area," and to report back to Congress with recommendations concerning future management of the area. Another section of the legislation provides that drilling cannot proceed in the 1002 Area until Congress authorizes it.

In 1998, the U.S. Geological Survey (USGS) estimated that the 1002 Area contained between 5.7 billion and 16 billion barrels of recoverable oil, making it the single most promising field in the United States. The midrange estimate, 10.4 billion barrels, is almost as much oil as that produced from North America's largest field, Prudhoe Bay, also in Alaska. Peak production could range between 1 million and 1.3 million barrels a day, which would represent more than 20 percent of the nation's oil production.

The administration of President George W. Bush supported drilling in ANWR and repeatedly clashed with Congress over the issue. In 2001, the National Energy Policy Development Group,* a panel created by the president, argued that the environmental consequences would be less serious than opponents claim:

> A lengthy 1999 Department of Energy study examined the environmental benefits of new exploration and production technologies and concluded that "improvements over the past 40 years have dramatically reduced industry's footprint on the fragile tundra, have minimized waste produced, and have protected the land for resident and migrating wildlife." The same study concluded "it is important to tell this remarkable story of environmental progress in E&P [exploration and production] technology. Greater

$11 billion to subsidize wind, solar, biomass, and geothermal power, with little to show for it. According to Schwartz and Reiss: "Despite all the hype, tax breaks, and incentives, the proportion

awareness of the industry's achievements in environmental protection will provide the context for effective policy, and for informed decision-making by both the private and public sectors."

It added:

> These [technological] advances include the use of ice roads and drilling pads, low-impact exploration approaches such as winter-only exploration activities, and extended reach and through tubing rotary drilling. These technologies have significantly reduced the size of production-related facilities on the North Slope.

The panel also concluded that no more than 2,000 acres would be disturbed if the 1002 Area of ANWR were developed. It said that, although ANWR is about the size of South Carolina, the developed area would be less than one-fifth the size of Dulles International Airport outside Washington, D.C.

Environmental groups disagree with the panel. They contend that allowing drilling in ANWR would have serious environmental consequences. According to the Sierra Club,** ANWR is one of America's last remaining wildernesses, and the 1002 Area is its "biological heart": It supports more than 200 species of animals, including caribou, grizzly bears, and migratory birds. The organization points out that, elsewhere in Alaska, oil drilling has resulted in hundreds of oil spills and leakage of toxic chemicals. It also disputes the National Energy Policy Development Group's contention that drilling would affect only 2,000 acres of ANWR. In fact, oil companies would have to build roads, pipelines, and other infrastructure to support the drilling throughout the 1002 Area, and that network of industrial sites would remain standing in the refuge after the oil is gone. In any event, the Sierra Club maintains that opening ANWR to drilling would only continue our dependence on oil. It also argues that the higher miles-per-gallon standards, required by energy legislation passed in 2007, will save more oil than might ever be produced from the ANWR.

* National Energy Policy Development Group. *National Energy Policy: Reliable, Affordable, and Environment Sound Energy for America's Future*. Washington, D.C., 2001.

** Sierra Club, "Chill the Drills! Myth vs. Fact in the Arctic National Wildlife Refuge: Don't Believe the Lies!" http://www.sierraclub.org/arctic/myths.

of US electricity production from renewables has actually fallen in the past 15 years, from 11.0 percent to 9.1 percent. That decline would be even worse without hydropower, which accounts for 92 percent of the world's renewable electricity."[15]

Free-market advocates insist that alternatives simply cannot compete. For example, in 1995, federal regulators overturned a requirement that utilities in California buy wind-generated power. The following year, Kenetech, the world's leading producer of wind-generated electricity, filed for bankruptcy. H. Sterling Burnett of the National Center for Policy Analysis explains why companies such as Kenetech could not survive:

> While the price of wind power has indeed fallen, it still costs more than spot market electric power (3.5 to 4 cents kwh). Furthermore, the price gap between wind and conventional power production is actually greater, since the federal government subsidizes wind power through a production tax credit of 1.8 cents per kwh. Wind power plants also receive accelerated depreciation, allowing owners to write off their costs in five years rather than the usual 20. . . . Thus, when the 1.8 cent kwh tax credit lapsed in 2003, new wind power projects suddenly became uncompetitive.[16]

Researchers at the Cato Institute raise another objection to subsidies for alternative energy. They point out that it is a "mature" industry, not a start-up, and one that has yet to develop competitive products despite years of effort.

Finally, subsidies cannot force technological changes to happen. One notable failure was California's attempt to promote electric cars: "Even massive tax incentives and manufacturers' subsidies didn't make much difference. The Toyota RAV I, for example, sold for over $40,000—more than double the price of a conventional gasoline RAV. Though consumers could claim $13,000 in state and federal tax credits for buying that model, there were still very few takers."[17] So few electric cars were sold that California scrapped its ambitious plan to require "zero-emissions" vehicles to account for 10 percent of new car sales by 2003. Another failed energy initiative was a 1978 federal law that required utility companies to connect their grids to the operators of wind farms and trash plants, and to buy the energy they generated—at premium prices. By one estimate, that law forced utility companies to spend $40 billion, for which they received a meager amount of electricity. Nevertheless, similar proposals are back on the table today despite the fact that many people are voluntarily seeking out economically viable ways of reducing dependence on fossil fuels—by purchasing, for example, compact fluorescents, hybrid vehicles, and energy-efficient windows—without government intervention.

Summary

For many years, government considered energy too vital to be controlled by market forces alone. Experts now consider that approach a mistake. We learned from the oil shocks of the 1970s that regulation makes energy less available. Regulators also lack the expertise to set energy policy. One example was California's electric-power crisis, which resulted from bad decisions by state officials. Heavy-handed regulation, such as a requirement that electric utilities generate part of their power from renewable fuels, drives up the cost of power. At the same time, subsidies for alternative energy waste tax dollars on products that are still uncompetitive with traditional fuels. In recent years, the command-and-control model has given way to deregulation, a policy that has spurred innovation in other industries and is likely to do so with energy as well.

Addressing Energy Depletion

Like other products and technologies we rely on, oil can be either a blessing or a curse. Professor Michael Klare explains:

> In its most desirable states, [petroleum] is a free-flowing liquid material you can pump from the ground, transport over long distances, and refine into fuels and commodities. This is the petroleum that propelled the rise of modern industrial society, which exists in the popular imagination as a source of mobility, agility, and freedom: the winged stallion of the old Mobil emblem, the leaping tiger of Exxon. But petroleum can also take the form of a dense, dark, viscous material, as it does in the famous tar pits of La Brea, in central Los Angeles. In this form, it is the very opposite of freedom: It can entrap you, engulf you, kill you.[1]

Two Threats: Peak Oil and Global Warming

We face two related threats as the result of consuming fossil fuels. One is peak oil. The authors of the Hirsch Report said: "The bottom line is that no one knows with certainty when world oil production will reach a peak, but geologists have no doubt that it will happen."[2] The other is global warming. The Intergovernmental Panel on Climate Change has concluded that fossil fuels give off greenhouse gases that help raise the Earth's temperature, and it warns that, if the world continues burning these fuels at the current rate, we run the risk of unstoppable climate change. Even though the United States has refused to sign the Kyoto Protocol, lawmakers face considerable pressure to address global warming. Observers believe that Congress will eventually pass cap-and-trade legislation that would in effect penalize heavy emitters of greenhouse gases. (A possible alternative to cap-and-trade is a "carbon tax," which would be levied on fossil fuels in much the same way "sin taxes" are levied on alcohol and tobacco.) Supporters argue that these measures would put a price on emissions, and that cap-and-trade would also set measurable goals for reducing emissions. Opponents contend that these measures would raise the price of energy, which would hurt the economy.

The Ongoing Debate About Energy

Many arguments about energy policy echo those raised during the 1970s. In 1977, President Jimmy Carter told Americans: "Within 10 years we would not be able to import enough oil—from any country, at any acceptable price" and that "[i]f we fail to act soon, we will face an economic, social, and political crisis that will threaten our free institutions."[3] Many observers believe that Carter was too pessimistic about the supply of energy. Fewer than 10 years later, a worldwide glut of oil developed and OPEC found itself weakened and divided. Critics also contend that the president's regulatory-heavy approach was a failure, and that high energy prices of the 1970s, not government mandates, encouraged production as well as conservation.

Nevertheless, some measures that date back to the 1970s survive. One is the Strategic Petroleum Reserve (SPR), the world's largest emergency supply. Today, the SPR holds more than 700 million barrels, enough to last more than a month if the supply of oil to America is cut off. Oil has been drawn from it occasionally, including during the first Gulf War and after Hurricane Katrina. During the summer of 2008, former Energy Department official

THE LETTER OF THE LAW

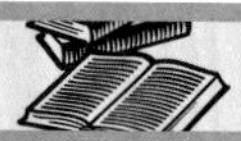

Renewable Portfolio Standards

During the past several decades, states abolished many of the regulations that govern utility companies. An exception to that trend is renewable portfolio standards (RPS) laws, which require that a certain percentage of a utility's power plant output come from renewable sources by a specific date.

Concern about global warming was a primary reason for RPS requirements. The requirement, however, was also motivated by a desire to achieve energy security, create "green" jobs, and make the air cleaner. In 2008, the American Wind Energy Association counted 25 states plus the District of Columbia with RPS laws and three other states with voluntary RPS targets.

In a typical RPS, the required percentage of renewables is 15 percent or 20 percent and the target date is 2015 or 2020. There is, however, considerable variation among the states. In addition, the definition of "renewable energy" varies by state; some states exempt municipal utilities; and others have different targets for large utilities than for small ones. Recently, a number of states have expanded their original RPS requirements. For example, Colorado's original target was 10 percent renewables by 2015, but lawmakers later increased it to 20 percent by 2020. Many states enforce RPS laws through a cap-and-trade system: Companies that are above the target can sell "credits" to companies that cannot meet the target. In addition, the federal government helps utilities meet RPS targets through policies such as production tax credits for renewable energy.

Nevada was one of the first states to adopt an RPS requirement. The original requirement, adopted in 1997, has been amended since then. The law (Sections

Joseph Romm called on the government to sell oil from the SPR. He told a U.S. House committee that doing so would provide more immediate relief than lifting restrictions on drilling:

> It is hard for me to see how anyone who thinks oil prices will drop if we end the federal moratorium on coastal drilling—which might deliver 100,000 barrels of oil a day sometime after 2020—could oppose releasing 500,000 barrels a day of oil starting now. Of course, the first strategy would benefit

704.7801 through 704.7828 of the *Nevada Revised Statutes*) defines "renewable energy" (Section 704.7811) to include biomass, geothermal energy, solar energy, wind, and water power (other than water power from Hoover Dam), and it specifically excludes fossil fuels and nuclear energy. Section 704.7821 sets out a schedule of RPS targets: 6 percent by 2005; 9 percent by 2007; 12 percent by 2009; 15 percent by 2011; 18 percent by 2013; and 20 percent by 2015. At least 5 percent of total power generated must be solar. A utility company can meet one-quarter of its RPS requirement by earning credits for "energy-efficiency measures," steps that reduce customers' energy use, that it can sell to companies that are unable to meet the RPS target. Section 704.7828 authorizes the Public Utilities Commission of Nevada to fine utilities that fail to comply with RPS requirements. A utility cannot pass the cost of fines onto its customers.

Supporters of RPS argue that the government must step in to increase the production of renewable energy to the point where it becomes competitive with energy from traditional sources. They believe that market forces alone will not bring about the needed transition from fossil fuels. Opponents argue that RPS standards enable regulators to second-guess the business judgment of utility companies and drive up electric bills by forcing utilities to use more expensive sources of power.

At the federal level, RPS legislation has passed the Senate several times since 2002 but has not won approval in the House. The House version of the 2007 energy legislation contained a 15-percent national RPS, but that requirement was taken out in the Senate when the bill failed to win the 60 votes needed to end a filibuster.

> oil companies and the second strategy would benefit the American people, so that may explain who supports which policy.[4]

The National Energy Policy Development Group, however, reported in 2001 that "the SPR is designed for addressing an imminent or actual disruption in oil supplies, and not for managing prices."[5]

CAFE standards also date back to the 1970s. They require an automaker's fleet of new cars to meet minimum fuel-economy standards. Joan Claybrook, the head of Public Citizen, told a U.S. Senate committee in 2002: "CAFE currently saves us 118 million gallons of gasoline every day and 913 million barrels of oil each year, or about the total imported annually from the Persian Gulf. It was a major factor in breaking the stranglehold of the Organization of the Petroleum Exporting Countries on oil prices and cutting rampant U.S. inflation in the early 1980s."[6] Still, many Americans oppose CAFE. Professors David Laband and Christopher Westley argue that the price of gas at the pump, not mileage standards, changes individuals' driving behavior. They note that, in 1980, the price of gas rose to nearly $3 a gallon in today's money. "People responded to that dramatic price increase by changing their behavior in ways that reduced the overall cost of driving: they carpooled more; they planned their shopping more carefully to reduce the number of driving trips taken; and they bought more fuel-efficient cars."[7] Laband and Westley add that CAFE standards restrict consumer choice and force some families to buy smaller cars that increase the risk of their being killed in a traffic crash.

Energy Policy in the Twenty-First Century

President Ronald Reagan, who defeated Jimmy Carter in the 1980 election, took an entirely different approach to energy policy. His administration's emphasis was on "deregulation," relying on market forces to encourage businesses to find energy and consumers to use it more efficiently. Since that time, deregulation has been the centerpiece of American energy policy, with conservation measures taking a low priority, especially after oil prices fell. The administration of President George W. Bush has emphasized a supply-side approach, which the National Energy Policy Development Group summed up as follows: "The United States supports a practical, market-based approach that encourages the adoption of efficient technologies, including those relating to natural gas, nuclear energy, and renewable energy."[8] The panel's recommendations focused on incentives to increase production and eliminating regulations that impede the delivery of energy.

Critics argue that the Bush administration's energy policy will make this country *more* dependent on fossil fuels. Some add that it not only favors the oil and gas industry but amounts to a payback for its contributions to the Republican Party. In addition, critics accuse the administration of ignoring conservation and turning a blind eye to the problem of energy depletion—even to the point of failing to plan for the peaking of oil production. Supporters of the Bush administration's policies counter that many restrictions on energy production are out of date. One example is the federal ban on oil exploration in Alaska. They cite the National Energy Policy Development Group's findings: "Technological improvements over the past 40 years have dramatically reduced industry's footprint on the tundra, minimized waste produced, and protected the land for resident and migratory wildlife. These advances include the use of ice roads and drilling pads, low-impact exploration approaches such as winter-only exploration activities, and extended reach and through tubing rotary drilling."[9] Supporters also note that the administration spent more than $10 billion to develop alternative energy sources, the Energy Department proposed higher CAFE standards for light trucks, the president announced a $1.2 billion Hydrogen Fuel Initiative, and the government is implementing legislation that requires the use of 7.5 billion gallons of ethanol and biodiesel fuel by 2012.

The September 11, 2001, terrorist attacks focused attention on America's energy security. Some commentators, such as *New York Times* columnist Thomas Friedman, argued that the United States should launch a national energy-independence program on the scale of the Apollo space program that put men on the Moon in the late 1960s and early 1970s. Paul Roberts describes what happened instead: "For Bush, the lesson to be learned about energy insecurity was not that the West should use less

The Apollo Alliance's Plan for Energy Independence

Numerous organizations have offered proposals for this country's energy policy. One of them is the Apollo Alliance, a coalition dedicated to bringing about a "clean energy revolution" that will end our dependence on foreign oil and reduce greenhouse-gas emissions. To achieve those goals, the alliance offers a 10-point plan:*

1. Promote advanced technology and hybrid cars by giving America's automakers financial aid to make the transition to advanced fuel-efficient cars.
2. Give tax breaks and economic development funds to manufacturers to make their factories more energy efficient.
3. Encourage the construction of "green" businesses and energy-efficient homes and offices through innovative financing and stronger building codes.
4. Provide American manufacturers incentives to make highly efficient goods and encourage consumers to buy them.
5. Develop an electric-power infrastructure that supports distributed generation (a network of small local production facilities rather than large power plants), and the use of renewables; apply the best available technology to reduce pollution at power plants; and support research into new

energy, as it did in the early 1980s, but that the West should be willing to make energy more secure and less unpredictable, as America had tried to do during the first Gulf War."[10]

Energy became a key issue in the 2008 election. The major political parties differ in their approaches to energy. Generally, Republicans prefer the Bush administration's supply-side approach, whereas Democrats emphasize conservation. Their differences came to a head over the issue of allowing oil and gas exploration in the Outer Continental Shelf and the Arctic. Republicans argued that the United States should do everything possible to exploit its own natural resources, and that energy security takes precedence over protecting the environment.

technology that will capture carbon dioxide emissions and store them where they cannot enter the atmosphere.

6. Promote existing technologies in solar, biomass, and wind energy; set "ambitious but achievable" goals for increasing the amount of energy generated by renewable fuels; and promote state and local policies that create "green" jobs.
7. Invest in alternatives to cars, including bicycle travel, local buses, light rail, regional high-speed railroads, and magnetic-levitation train projects.
8. Revitalize urban areas by rebuilding and upgrading roads, bridges, and water and wastewater systems; expand the redevelopment of abandoned urban "brownfields"; and improve metropolitan planning and government.
9. Invest in long-term research and development of hydrogen fuel cell technology, and build the infrastructure to support hydrogen-powered cars and distributed electricity generation using fuel cells as fixed locations.
10. Adopt regulations that promote diverse sources of energy and a reliable system for delivering energy, reward consumers for conservation efforts, and give emerging technologies a chance to succeed.

* Apollo Alliance, "The Ten-Point Plan for Good Jobs and Energy Independence." http://www.apolloalliance.org/resources_tenpointplan.php.

Democrats responded that opening all of our oil and gas reserves to drilling would only slightly reduce our reliance on foreign suppliers while at the same time continue to pollute the environment. They argue that increased conservation will do more in the long run to achieve energy independence.

Lawmakers React to the Energy Crisis

In 2005, Congress passed the Energy Policy Act,[11] which offered a variety of incentives to energy producers. Critics argued that it was a "giveaway" to the energy companies that did little to reduce demand or encourage the use of alternatives. Two years later, lawmakers passed the Energy Independence and Security Act of 2007.[12] Among other mandates, it increases the CAFE standard on automobiles to 35 mpg by 2020. The legislation also directs the Transportation Department to develop a system that rates cars on the basis of fuel economy and greenhouse-gas emissions, and provides funding for alternative technology such as hybrid vehicles, which can run on either gasoline or electricity. Critics fault the legislation for subsidizing the production of corn-based ethanol, which has been blamed for environmental damage and higher food prices.

Half the states in the Union have passed renewable portfolio standards (RPS) laws, which require utilities to derive a specific percentage of their electric power from renewable sources. The typical RPS requirement calls for 15 percent to 20 percent renewables and sets 2015 or 2020 as the target for reaching that goal. At the federal level, RPS legislation has been introduced several times but has yet to win approval from Congress.

Other proposals have been offered as well. One is the "feebate," which would tax energy-inefficient cars and distribute the revenue to buyers of fuel-efficient models. Roberts explains: "The feebate logic is simple: when buying a car, American consumers pay far more attention to up-front costs, such as sticker price, than to 'life-cycle' costs, such as fuel or maintenance. Adding five thousand dollars to the sticker price, auto industry

Pushing renewable energy

States with standards to require or encourage utility companies to increase their share of electricity produced from renewable sources:

	Percent renewable	Reached by
Ariz.	15%	2025
Calif.	20%	2010
Colo.	10%	2015
Conn.	10%	2010
D.C.	11%	2022
Del.	10%	2019
Hawaii	20%	2020
Iowa	2%	1999
Ill.	8%	2013
Maine	30%	2000
Md.	7.5%	2019
Mass.	4%	2009
Minn.	19%	2015
Mont.	15%	2015
N.M.	10%	2011
N.Y.	24%	2013
N.J.	22.5%	2020
Nev.	20%	2015
Pa.	8%	2020
R.I.	16%	2019
Texas	5.5%	2015
Wash.	15%	2020
Wis.	10%	2015

R.I.
D.C.
Conn.
Mass.
Del.
N.J.

Source: Union of Concerned Scientists (U.S.)
Graphic: Melina Yingling, Judy Treible

A map of the United States shows the 23 states (as of 2006) with standards that require or encourage energy companies to produce a share of the state's electricity using renewable energy sources.

analysts say, would be enough to persuade most buyers to take a look at more efficient models."[13] Opponents contend out that gas prices themselves provide incentive enough to conserve. They remind us that demand for oil fell more than 10 percent after the 1979 crisis and took 17 years to recover.

Another proposal is net metering, which requires utility companies to buy energy that their customers produce from sources such as solar power. Several states, including Colorado and New Jersey, have passed versions of net-metering laws. Supporters say that net metering would encourage small, decentralized producers. Opponents argue that net metering, like RPS requirements, would increase customers' utility bills and subsidize uncompetitive producers.

Looking Ahead

Many experts believe that natural gas could provide a short-term "bridge" between our current dependence on oil and alternative fuels of the future. In his energy plan,[14] T. Boone Pickens argues that we should build cars that run on natural gas. He points

The Energy Policies Proposed by Barack Obama During the 2008 Presidential Campaign

A surge in energy prices coincided with the 2008 presidential election, making energy an important concern of voters. In many respects, the two major parties offered differing approaches to energy policy. Traditionally, Republicans prefer policies that increase the supply of energy and believe that market forces will lead to a solution to the nation's energy problem, whereas Democrats emphasize the importance of conservation and believe that government should play a leading role in the transition from fossil fuels to renewable energy sources.

During his successful run for the presidency in 2008, Barack Obama, the Democratic Party's nominee for president, proposed:

- Providing an emergency energy rebate of up to $1,000 per household and expanding weatherization assistance to low-income households.
- Releasing some of the oil from the Strategic Petroleum Reserve to help bring down the price.
- Investing $150 billion during the next 10 years to help develop alternative forms of energy and create "green" jobs.
- Putting 1 million plug-in hybrid cars on the road by 2015.
- Increasing CAFE standards by 4 percent a year and providing $4 billion in subsidies for the American auto industry to help build more fuel-efficient cars.

out that 98 percent of the natural gas we use comes from the United States and Canada. Yet despite its abundance, we use only 1 percent of that gas in automobiles. Natural gas, however, has drawbacks of its own. Like oil, the supply is finite, and it emits greenhouse gases when burned. Furthermore, experts believe that North American production is peaking, which means that we could find ourselves depending on imports from many of the same countries that control much of the world's oil.

- Imposing tougher energy-efficiency standards for buildings and appliances.
- Promoting the responsible domestic production of oil and natural gas, and safe and secure nuclear energy.

Obama agreed in principle with Senator John McCain, his Republican rival for the presidency in 2008, on the following measures:

- A crackdown on oil speculators.
- Requiring new vehicles to incorporate "flex-fuel" technology, which allows them to run on alcohol fuels as well as gasoline.
- Making the federal government a leader in energy conservation.
- "Smart metering," which gives consumers an incentive to use electric power more wisely.
- A "cap-and-trade" system to limit greenhouse-gas emissions. Under that approach, the government places a limit, or cap, on emissions; companies that stay under the limit may sell credits for the remaining emissions to companies that are over the limit. This proposal is aimed at curbing global warming, but it also has an effect on the price of energy and the future mix of energy sources.

Sources: http://www.johnmccain.com; http://www.barackobama.com.

Some people envision a day when America will have a "hydrogen economy" where energy is cheap, decentralized, and nonpolluting. According to Peter Huber and Mark Mills:

> If solar-electric power is used to extract hydrogen from water, there will be no more need for big central power plants and filling stations. We will decommission the nuclear power plants, shut down the strip mines, and scrap the offshore oil wells. Silicon and hydrogen will completely displace uranium and carbon. We will be all-solar and all-electric. Silicon, wind, and sun—earth, air, and fire—will move the fourth element, water, to power every aspect of our lives.[15]

Others maintain that, no matter how much money is spent, hydrogen technology will never become practical. One of them is Enoch Durbin, a professor at Princeton University. He argues:

> To make hydrogen available for use as an energy source, we have to separate it from the elements to which it is bound, such as water (H_2O) or fossil fuels. This is somewhat like converting ashes of coal back into coal. . . .
>
> Once hydrogen is made it must be shipped to where it is needed. Distributing hydrogen is another problem. Typically gas or liquid energy is distributed by use of a pipeline. We do not have the pipelines to distribute hydrogen and they are extremely difficult to create. Electrical energy is transmitted over wires, a much easier distribution system. Why then would anyone go to all the trouble of making, storing, and shipping hydrogen?[16]

How will the current energy crisis play out? Dr. Michael Smith, an energy expert from Britain, believes we are on the verge of serious disruption. He recently warned: "[T]here is no painless way to fill the gap. Of course it will be filled, partly from traditional sources, partly from new alternatives, partly from simple efficiencies, but a large portion will have to be filled by demand destruction. In the real world demand destruction means poverty and conflict so we should be working towards reducing our vulnerability to such destruction."[17] On the other hand, U.S. Representative Roscoe Bartlett, who formed the House Peak Oil Caucus, sounds an optimistic note. He said: "This is a huge challenge. We have the most innovative, creative society in the world. Properly informed and properly motivated, I think we're equal to the challenge. I see this as a very challenging fun future, where we really have something we can all pull together to accomplish."[18]

Summary

America faces two threats, both of which stem from our use of fossil fuels. Oil production will inevitably peak, though we are not sure when the peak will arrive. At the same time, there is wide agreement that the burning of fossil fuels releases greenhouse gases that contribute to global warming. To meet these threats, Americans will have to make drastic cuts in their fossil-fuel consumption. At the federal level, deregulation dominates our energy policy. Congress has taken modest steps to move the nation away from fossil fuels, but federal policy is still tilted toward production of those fuels and gives conservation a low priority. Some states have passed laws aimed at reducing consumption and encouraging the production of energy from alternative fuels. The debate continues about whether government should play a greater role in setting energy policy and which alternatives will replace today's fuels.

Beginning Legal Research

The goals of each book in the POINT/COUNTERPOINT series are not only to give the reader a basic introduction to a controversial issue affecting society, but also to encourage the reader to explore the issue more fully. This Appendix is meant to serve as a guide to the reader in researching the current state of the law as well as exploring some of the public policy arguments as to why existing laws should be changed or new laws are needed.

Although some sources of law can be found primarily in law libraries, legal research has become much faster and more accessible with the advent of the Internet. This Appendix discusses some of the best starting points for free access to laws and court decisions, but surfing the Web will uncover endless additional sources of information. Before you can research the law, however, you must have a basic understanding of the American legal system.

The most important source of law in the United States is the Constitution. Originally enacted in 1787, the Constitution outlines the structure of our federal government, as well as setting limits on the types of laws that the federal government and state governments can enact. Through the centuries, a number of amendments have added to or changed the Constitution, most notably the first 10 amendments, which collectively are known as the "Bill of Rights" and which guarantee important civil liberties.

Reading the plain text of the Constitution provides little information. For example, the Constitution prohibits "unreasonable searches and seizures" by the police. To understand concepts in the Constitution, it is necessary to look to the decisions of the U.S. Supreme Court, which has the ultimate authority in interpreting the meaning of the Constitution. For example, the U.S. Supreme Court's 2001 decision in *Kyllo v. United States* held that scanning the outside of a person's house using a heat sensor to determine whether the person is growing marijuana is an unreasonable search—if it is done without first getting a search warrant from a judge. Each state also has its own constitution and a supreme court that is the ultimate authority on its meaning.

Also important are the written laws, or "statutes," passed by the U.S. Congress and the individual state legislatures. As with constitutional provisions, the U.S. Supreme Court and the state supreme courts are the ultimate authorities in interpreting the meaning of federal and state laws, respectively. However, the U.S. Supreme Court might find that a state law violates the U.S. Constitution, and a state supreme court might find that a state law violates either the state or U.S. Constitution.

Not every controversy reaches either the U.S. Supreme Court or the state supreme courts, however. Therefore, the decisions of other courts are also important. Trial courts hear evidence from both sides and make a decision, while appeals courts review the decisions made by trial courts. Sometimes rulings from appeals courts are appealed further to the U.S. Supreme Court or the state supreme courts.

Lawyers and courts refer to statutes and court decisions through a formal system of citations. Use of these citations reveals which court made the decision or which legislature passed the statute, and allows one to quickly locate the statute or court case online or in a law library. For example, the Supreme Court case *Brown v. Board of Education* has the legal citation 347 U.S. 483 (1954). At a law library, this 1954 decision can be found on page 483 of volume 347 of the U.S. Reports, which are the official collection of the Supreme Court's decisions. On the following page, you will find sample of all the major kinds of legal citation.

Finding sources of legal information on the Internet is relatively simple thanks to "portal" sites such as findlaw.com and lexisone.com, which allow the user to access a variety of constitutions, statutes, court opinions, law review articles, news articles, and other useful sources of information. For example, findlaw.com offers access to all Supreme Court decisions since 1893. Other useful sources of information include gpo.gov, which contains a complete copy of the U.S. Code, and thomas.loc.gov, which offers access to bills pending before Congress, as well as recently passed laws. Of course, the Internet changes every second of every day, so it is best to do some independent searching.

Of course, many people still do their research at law libraries, some of which are open to the public. For example, some state governments and universities offer the public access to their law collections. Law librarians can be of great assistance, as even experienced attorneys need help with legal research from time to time.

Common Citation Forms

Source of Law	Sample Citation	Notes
U.S. Supreme Court	*Employment Division v. Smith*, 485 U.S. 660 (1988)	The U.S. Reports is the official record of Supreme Court decisions. There is also an unofficial Supreme Court ("S. Ct.") reporter.
U.S. Court of Appeals	*United States v. Lambert*, 695 F.2d 536 (11th Cir.1983)	Appellate cases appear in the Federal Reporter, designated by "F." The 11th Circuit has jurisdiction in Alabama, Florida, and Georgia.
U.S. District Court	*Carillon Importers, Ltd. v. Frank Pesce Group, Inc.*, 913 F.Supp. 1559 (S.D.Fla.1996)	Federal trial-level decisions are reported in the Federal Supplement ("F. Supp."). Some states have multiple federal districts; this case originated in the Southern District of Florida.
U.S. Code	Thomas Jefferson Commemoration Commission Act, 36 U.S.C., §149 (2002)	Sometimes the popular names of legislation—names with which the public may be familiar—are included with the U.S. Code citation.
State Supreme Court	*Sterling v. Cupp*, 290 Ore. 611, 614, 625 P.2d 123, 126 (1981)	The Oregon Supreme Court decision is reported in both the state's reporter and the Pacific regional reporter.
State Statute	Pennsylvania Abortion Control Act of 1982, 18 Pa. Cons. Stat. 3203-3220 (1990)	States use many different citation formats for their statutes.

Cases and Statutes

The Energy Policy and Conservation Act of 1975 (EPCA) (Public Law 94–163)
The first comprehensive energy legislation passed after the Arab oil embargo. One key provision was the establishment of Corporate Average Fuel Economy (CAFE) standards for automobiles and most light trucks. Congress has passed a number of energy-related bills since then.

The Energy Policy Act of 2005 (Public Law 109–58)
A package of incentives for the production of traditional and alternative fuels.

The Energy Independence and Security Act of 2007 (Public Law 110–140)
This act is more conservation oriented than the 2005 legislation. One provision of the law will increase CAFE's fuel-efficiency standard by 10 miles per gallon by 2020.

***Center for Biological Diversity v. National Highway Traffic Safety Administration*, 508 F.3d 550 (9th Cir. 2007)**
A significant case in energy policy because it was a legal challenge to the government over CAFE's fuel-efficiency standards, which activists argued were too weak.

Terms and Concepts

Alternative fuels
Arab oil embargo
Arctic National Wildlife Refuge
Biomass
Clean coal
Conservation
Corporate Average Fuel Economy (CAFE) standards
Department of Energy
Deregulation
Distributed networks
Energy efficiency
Energy independence
Ethanol
External costs
"Feebate"
Fossil fuels
Greenhouse gases
Hirsch Report
Hybrid car

Hydrogen economy
Infrastructure
Internal-combustion engine
Leadership in Energy and Environmental Design (LEED)
"Moral equivalent of war"
National Energy Policy Development Group
Natural gas
Net metering
Nuclear power
Organization of the Petroleum Exporting Countries (OPEC)
Peak oil
Proved/proven reserves
Renewable energy sources
Renewable portfolio standards
Strategic Petroleum Reserve
Unconventional oil

NOTES

Introduction: Energy and the Industrialized World

1 Vijay V. Vaitheeswaran, *Power to the People: How the Coming Energy Revolution Will Transform an Industry, Change Our Lives, and Maybe Even Save the Planet*. New York: Farrar, Strauss and Giroux, 2003, pp. 183–184.
2 M. King Hubbert, "Nuclear Energy and the Fossil Fuels." Presented before the Spring Meeting of the Southern District Division of Production, American Petroleum Institute, March 7–9, 1956. Houston: Shell Development Company, 1956.
3 Michael T. Klare, *Blood for Oil: The Dangers and Consequences of America's Growing Dependency on Imported Petroleum*. New York: Metropolitan Books, 2004, p. 8.
4 Vaitheeswaran, *Power to the People*, p. 265.
5 Kenneth S. Deffeyes, *Beyond Oil: The View From Hubbert's Peak*. New York: Hill and Wang, 2005, p. 5.
6 Jimmy Carter, "The President's Proposed Energy Policy," Address to the American People, April 18, 1977. Published in *Vital Speeches of the Day* 43, no. 14 (May 1, 1977), pp. 418–420.
7 Jimmy Carter, State of the Union Address, January 23, 1980. http://www.jimmycarterlibrary.org/documents/speeches/su80jec.phtml.
8 National Energy Policy Development Group, "National Energy Policy: Reliable, Affordable, and Environment Sound Energy for America's Future," Washington, D.C., 2001, p. viii.
9 U.N. Intergovernmental Panel on Climate Change, "Fourth Assessment Report. Climate Change 2007: Synthesis Report. Summary for Policymakers," Geneva, Switzerland, 2007, p. 6.
10 Hubbert, "Nuclear Energy and the Fossil Fuels."

Point: Energy Depletion Is a Serious Threat

1 Paul Roberts, *The End of Oil: On the Edge of a Perilous New World*. Boston: Houghton Mifflin, 2004, p. 44.
2 Oil Depletion Analysis Centre, *Preparing for Peak Oil*. London, 2007, p. 2.
3 M. King Hubbert, *Nuclear Energy and the Fossil Fuels*. Presented before the Spring Meeting of the Southern District Division of Production, American Petroleum Institute, March 7–9, 1956. Houston: Shell Development Company, 1956.
4 Robert L. Hirsch, Roger Bezdek, and Robert Wendling, *Peaking of World Oil Production: Impacts, Mitigation & Risk Management*. San Diego: Science Applications International Corporation, 2005, p. 36.
5 Roberts, *The End of Oil*, p. 245.
6 Hirsch, Bezdek, and Wendling, *Peaking of World Oil Production*, p. 36.
7 Roscoe Bartlett, "Peak Oil Address to the House of Representatives," *Congressional Record*, February 7, 2008.
8 Kenneth S. Deffeyes, *Beyond Oil: The View From Hubbert's Peak*. New York: Hill and Wang, 2005, p. 182.
9 Colin Campbell, "Peak Oil—A Turning Point for Mankind," *Social Contract* 15, no. 3 (Spring 2005).
10 Union of Concerned Scientists, *Energy Security: Solutions to Protect America's Power Supply and Reduce Oil Dependence*. Cambridge, Mass.: Union of Concerned Scientists, 2002, p. 2.
11 Paul Roberts, "Over a Barrel," *Mother Jones* 29 (November/December 2004), p. 64.
12 Roberts, *The End of Oil*, p. 13.
13 James Howard Kunstler, *The Long Emergency: Surviving the Converging Catastrophes of the Twenty-First Century*. New York: Atlantic Monthly Press, 2005, p. 65.
14 Hirsch, Bezdek, and Wendling, *Peaking of World Oil Production*, p. 64.
15 Deffeyes, *Beyond Oil*, p. 179.
16 Hirsch, Bezdek, and Wendling, *Peaking of World Oil Production*, p. 59.
17 Roberts, *The End of Oil*, p. 306.
18 Hirsch, Bezdek, and Wendling, *Peaking of World Oil Production*, p. 27.

Counterpoint: Energy Is Abundant and Will Remain So

1 Stephen Moore, "A Century of Environmental Progress and Natural Resource Abundance." In *Global Warming and Other Eco-Myths: How the Environmental Movement Uses False Science to Scare Us*

to Death, edited by Ronald Bailey, p. 96. Roseville, Calif.: Prima Publishing, 2002.

2 M.A. Adelman, "The Real Oil Problem," *Regulation* 27, no. 1 (Spring 2004), p. 18.

3 U.S. Government Accountability Office. *Uncertainty About Future Oil Supply Makes It Important to Develop a Strategy for Addressing a Peak and Decline in Oil Production*. Report no. GAO-07-283. Washington, D.C.: 2007, p. 17.

4 David Deming, *Are We Running Out of Oil? National Center for Policy Analysis Policy Backgrounder no. 159.* Washington, D.C., 2003, p. 10.

5 Paul Roberts, *The End of Oil: On the Edge of a Perilous New World*. Boston: Houghton Mifflin, 2004, p. 167.

6 Adelman, "The Real Oil Problem," pp. 16–17.

7 Deming, *Are We Running Out of Oil?*, pp. 7–8.

8 Peter Huber and Mark Mills, "Oil, Oil, Everywhere . . . ," *Wall Street Journal*, January 27, 2005.

9 *Ibid.*

10 Vijay V. Vaitheeswaran, *Power to the People: How the Coming Energy Revolution Will Transform an Industry, Change Our Lives, and Maybe Even Save the Planet*. New York: Farrar, Strauss and Giroux, 2003, pp. 105–106.

11 Huber and Mills, "Oil, Oil, Everywhere. . . ."

12 National Energy Policy Development Group, *National Energy Policy: Reliable, Affordable, and Environment Sound Energy for America's Future*. Washington, D.C., 2001, p. 71.

Point: Continued Reliance on Fossil Fuels Is Dangerous

1 Michael T. Klare, *Blood for Oil: The Dangers and Consequences of America's Growing Dependency on Imported Petroleum*. New York: Metropolitan Books, 2004, p. 36.

2 *Ibid.*, p. 66.

3 Joshua Kurlantzick, "Put a Tyrant in Your Tank," *Mother Jones* (May/June 2008).

4 Matthew Simmons, "Challenges in a World of Oil Scarcity: The Coming Saudi Oil Crisis," CounterPunch.com, June 21, 2005. http://www.counterpunch.org/simmons06212005.html.

5 Paul Roberts, *The End of Oil: On the Edge of a Perilous New World*. Boston: Houghton Mifflin, 2004, p. 102.

6 Lutz Kleveman, "Oil and the New Great Game," *Nation*, February 16, 2004.

7 Klare, *Blood for Oil*, p. 21.

8 Roberts, *The End of Oil*, p. 105.

9 Klare, *Blood for Oil*, p. 6.

10 *Ibid.*, pp. 161–162.

11 Michael L. Ross, "Blood Barrels: Why Oil Wealth Fuels Conflict," *Foreign Affairs* 87, no. 3 (May/June 2008), p. 2.

12 *Ibid.*

13 Robert L. Hirsch, Roger Bezdek, and Robert Wendling, *Peaking of World Oil Production: Impacts, Mitigation & Risk Management*. San Diego: Science Applications International Corporation, 2005, p. 62.

14 United Nations Intergovernmental Panel on Climate Change, *Fourth Assessment Report: Climate Change 2007: Synthesis Report. Summary for Policymakers.* Geneva, Switzerland, 2007, p. 30.

15 Oil Depletion Analysis Centre, *Preparing for Peak Oil*. London, 2007, p. 4.

16 Peter Schwartz and Spencer Reiss, "Nuclear Now!," *Wired* 13, no. 2 (February 2005).

Counterpoint: Alternative Energy Is Riskier Than Traditional Fuels

1 H. Sterling Burnett, "Wind Power: Red Not Green." Policy Analysis Brief Analysis No. 467. Washington, D.C.: National Center for Policy Analysis, 2004.

2 Peter Schwartz and Spencer Reiss, "Nuclear Now!," *Wired* 13, no. 2 (February 2005).

3 Paul Roberts, *The End of Oil: On the Edge of a Perilous New World*. Boston: Houghton Mifflin, 2004, p. 203.

4 David Morris, "A Hydrogen Economy is a Bad Idea," AlterNet.org, February 24, 2003. http://www.alternet.org/story/15239.

5 U.S. Government Accountability Office, "Uncertainty About Future Oil Supply Makes It Important to Develop

a Strategy for Addressing a Peak and Decline in Oil Production." Report no. GAO-07-283. Washington, D.C.: 2007, p. 30.

6 *Ibid.*

7 Mayer Hillman with Tina Fawcett and Sudhir Chella Rajan, *The Suicidal Planet: How to Prevent Global Climate Catastrophe*. New York: Thomas Dunne Books, 2007, p. 101.

8 Vijay V. Vaitheeswaran, *Power to the People: How the Coming Energy Revolution Will Transform an Industry, Change Our Lives, and Maybe Even Save the Planet*. New York: Farrar, Strauss and Giroux, 2003, p. 305.

9 U.S. Government Accountability Office, *Uncertainty About Future Oil Supply Makes It Important to Develop a Strategy for Addressing a Peak and Decline in Oil Production*, p. 20.

10 *Ibid.*

11 *Ibid.*

12 Brown is quoted in Oil Depletion Analysis Centre, *Preparing for Peak Oil*, 2007, p. 7.

13 George Monbiot, "Fuel for Nought," *Guardian*, November 23, 2004.

14 Peter W. Huber and Mark P. Mills, *The Bottomless Well: The Twilight of Fuel, the Virtue of Waste, and Why We Will Never Run Out of Energy*. New York: Basic Books, 2005.

15 Bjorn Lomborg, *Cool It: The Skeptical Environmentalist's Guide to Global Warming*. New York: Alfred A. Knopf, 2007, p. 155.

16 *Ibid.*, pp. 155–156.

17 Michael Totty, "The Case For and Against Nuclear Power," *Wall Street Journal*, June 30, 2008.

18 National Energy Policy Development Group, *National Energy Policy: Reliable, Affordable, and Environment Sound Energy for America's Future*. Washington, D.C., 2001, pp. 5–16.

19 "The Science of Climate Change," Senate Floor Statement by U.S. Senator James Inhofe, chairman, Committee on the Environment and Public Works, July 28, 2003.

20 *Ibid.*

Point: Government Must Address the Energy Crisis

1 Robert L. Hirsch, Roger Bezdek, and Robert Wendling, *Peaking of World Oil Production: Impacts, Mitigation & Risk Management*. San Diego: Science Applications International Corporation, 2005, p. 13.

2 Paul Roberts, "Over a Barrel," *Mother Jones* 29 (November/December 2004), p. 64.

3 Paul Roberts, *The End of Oil: On the Edge of a Perilous New World*. Boston: Houghton Mifflin, 2004, p. 230.

4 Ross Gelbspan, *Boiling Point: How Politicians, Big Oil and Coal, Journalists, and Activists Are Fueling the Climate Crisis—and What We Can Do to Avert Disaster*. New York: Basic Books, 2004, p. 133.

5 Roberts, *The End of Oil*, p. 219.

6 Hirsch, Bezdek, and Wendling, *Peaking of World Oil Production*, p. 66.

7 Roberts, *The End of Oil*, p. 15.

8 Mayer Hillman with Tina Fawcett and Sudhir Chella Rajan, *The Suicidal Planet: How to Prevent Global Climate Catastrophe*. New York: Thomas Dunne Books, 2007, p. 148.

9 Union of Concerned Scientists, *Energy Security: Solutions to Protect America's Power Supply and Reduce Oil Dependence*. Cambridge, Mass.: Union of Concerned Scientists, 2002, p. 10.

10 Al Gore, "A Generational Challenge to Repower America." Speech given in Washington, D.C., July 17, 2008. http://www.wecansolveit.org/pages/al_gore_a_generational_challenge_to_repower_america.

11 Vijay V. Vaitheeswaran, *Power to the People: How the Coming Energy Revolution Will Transform an Industry, Change Our Lives, and Maybe Even Save the Planet*. New York: Farrar, Strauss and Giroux, 2003, p. 251.

12 Public Law 94–163.

13 Craig Morris, "Renewables From the Bottom Up," Truthout.org, January 29, 2008. http://www.truthout.org/issues_06/012908EB.shtml.

14 Roberts, *The End of Oil*, p. 200.

Counterpoint: A Free Market Is the Soundest Energy Policy

1 Peter W. Huber and Mark P. Mills, *The Bottomless Well: The Twilight of Fuel, the Virtue of Waste, and Why We Will Never Run Out of Energy*. New York: Basic Books, 2005, p. 76.
2 *Ibid.*
3 Vijay V. Vaitheeswaran, *Power to the People: How the Coming Energy Revolution Will Transform an Industry, Change Our Lives, and Maybe Even Save the Planet*. New York: Farrar, Strauss and Giroux, 2003, pp. 46–47.
4 Huber and Mills, *The Bottomless Well*, p. 106.
5 Jerry Taylor, "Not Cheap, Not Green," *Washington Times*, August 4, 2003.
6 http://www.energystar.gov/index.cfm?c=about.ab_history.
7 Robert L. Hirsch, Roger Bezdek, and Robert Wendling, *Peaking of World Oil Production: Impacts, Mitigation & Risk Management*. San Diego: Science Applications International Corporation, 2005, p. 71.
8 National Energy Policy Development Group, *National Energy Policy: Reliable, Affordable, and Environment Sound Energy for America's Future*. Washington, D.C., 2001, pp. 5–12.
9 Peter Schwartz and Spencer Reiss, "Nuclear Now!," *Wired* 13, no. 2 (February 2005).
10 Thomas Sowell, "Electricity Shocks California," *Jewish World Review*, January 12, 2001.
11 Vaitheeswaran, *Power to the People*, p. 63.
12 Ed Pilkington, "Big Oil to Big Wind: Texas Veteran Sets Up $10bn Clean Energy Project," *Guardian*, April 14, 2008.
13 Paul Roberts, *The End of Oil: On the Edge of a Perilous New World*. Boston: Houghton Mifflin, 2004, p. 215.
14 Vaitheeswaran, *Power to the People*, p. 183.
15 Schwartz and Reiss, "Nuclear Now!"
16 H. Sterling Burnett, *Wind Power: Red Not Green*. Policy Analysis Brief Analysis No. 467. Washington, D.C.: National Center for Policy Analysis, 2004.
17 Vaitheeswaran, *Power to the People*, p. 192.

Conclusion: Addressing Energy Depletion

1 Michael T. Klare, *Blood for Oil: The Dangers and Consequences of America's Growing Dependency on Imported Petroleum*. New York: Metropolitan Books, 2004, p. 180.
2 Robert L. Hirsch, Roger Bezdek, and Robert Wendling, *Peaking of World Oil Production: Impacts, Mitigation & Risk Management*. San Diego: Science Applications International Corporation, 2005, p. 8.
3 Jimmy Carter, "The President's Proposed Energy Policy." Address to the American People, April 18, 1977. Published in *Vital Speeches of the Day* 43, no. 14 (May 1, 1977), pp. 418–420.
4 Testimony of Joseph Romm Before the House Select Committee on Energy Independence and Global Warming, July 23, 2008. http://www.americanprogressaction.org/issues/2008/romm_spr_testimony.html.
5 National Energy Policy Development Group, *National Energy Policy: Reliable, Affordable, and Environment Sound Energy for America's Future*. Washington, D.C., 2001, Appendix I, Summary of Recommendations, Chapter 8.
6 Joan Claybrook, Testimony Before the Senate Committee on Commerce, Science and Transportation, January 24, 2002. http://www.citizen.org/documents/CAFE_testimony.pdf.
7 David N. Laband and Christopher Westley, "How Not to Respond to Higher Gasoline Prices," *Freeman*, October 2004, pp. 23–24.
8 National Energy Policy Development Group, *National Energy Policy*, pp. 8–14 & 8–15.
9 *Ibid.*, pp. 5–9.
10 Paul Roberts, *The End of Oil: On the Edge of a Perilous New World*. Boston: Houghton Mifflin, 2004, p. 112.
11 Public Law 109–58.
12 Public Law 110–140.

13 Roberts, *The End of Oil*, p. 318.
14 T. Boone Pickens. "The Pickens Plan for Energy Independence." Dallas, Tex., 2008. http://media.pickensplan.com/pdf/pickensplan.pdf.
15 Peter W. Huber and Mark P. Mills, *The Bottomless Well: The Twilight of Fuel, the Virtue of Waste, and Why We Will Never Run Out of Energy*. New York: Basic Books, 2005, p. 87.
16 Professor Durbin is quoted in Vijay V. Vaitheeswaran, *Power to the People: How the Coming Energy Revolution Will Transform an Industry, Change Our Lives, and Maybe Even Save the Planet.* New York: Farrar, Strauss and Giroux, 2003, pp. 241–242.
17 Michael R. Smith, "Is There a Painless Way to Fill the Oil Supply Gap?" EnergyFiles.com. http://www.energyfiles.com/oilsupplygap.html.
18 Roscoe Bartlett, Peak Oil Address to the House of Representatives, *Congressional Record*, February 7, 2008.

RESOURCES

Books

Gore, Al. *An Inconvenient Truth: The Planetary Emergency of Global Warming and What We Can Do About It.* Emmaus, Pa.: Rodale Press, 2006.

Huber, Peter W., and Mark P. Mills. *The Bottomless Well: The Twilight of Fuel, the Virtue of Waste, and Why We Will Never Run Out of Energy.* New York: Basic Books, 2005.

Klare, Michael T. *Blood for Oil: The Dangers and Consequences of America's Growing Dependency on Imported Petroleum.* New York: Metropolitan Books, 2004.

Roberts, Paul. *The End of Oil: On the Edge of a Perilous New World.* Boston: Houghton Mifflin, 2004.

Vaitheeswaran, Vijay V. *Power to the People: How the Coming Energy Revolution Will Transform an Industry, Change Our Lives, and Maybe Even Save the Planet.* New York: Farrar, Strauss and Giroux, 2003.

Reports

Hirsch, Robert L., Roger Bezdek, and Robert Wendling, *Peaking of World Oil Production: Impacts, Mitigation & Risk Management.* San Diego: Science Applications International Corporation, 2005.

National Energy Policy Development Group. *National Energy Policy: Reliable, Affordable, and Environment Sound Energy for America's Future.* Washington, D.C., 2001.

United Nations Intergovernmental Panel on Climate Change. *Fourth Assessment Report. Climate Change 2007: Synthesis Report. Summary for Policymakers.* Geneva, Switzerland: Intergovernmental Panel on Climate Change, 2007.

Web Sites

Apollo Alliance

http://www.apolloalliance.org

This organization offers a comprehensive plan to address the energy problems facing the United States.

BP

http://www.bp.com

The energy company BP, formerly known as British Petroleum, publishes the *BP Statistical Review of World Energy*, which is considered authoritative.

Energy Information Administration

http://www.eia.doe.gov

The EIA is the statistics-gathering arm of the U.S. Department of Energy.

Environmental Protection Agency

http://www.epa.gov

Energy policy often raises environmental concerns, and the lead federal agency in that area is the Environmental Protection Agency.

National Center for Policy Analysis

http://www.ncpa.org

This public policy institute believes that market forces, not government programs and regulations, will lead to solutions to our energy problems.

Natural Resources Defense Council

http://www.nrdc.org

This organization is urging the U.S. government to take steps to reduce our fossil-fuel consumption, not only to end our dependence on foreign oil and avoid the consequences of energy depletion but also to stop global warming.

Oil Depletion Analysis Centre

http://www.odac-info.org

This British-based organization was formed to raise public awareness about peak oil.

Union of Concerned Scientists

http://www.ucsusa.org

The Union of Concerned Scientists, a nonprofit science advocacy group based in the United States, has offered a comprehensive plan to address the nation's energy problems.

U.S. Department of Energy

http://www.doe.gov

This department is responsible for ensuring the nation's energy security and encouraging new energy-related technology.

U.S. Geological Service

http://www.usgs.gov

A division of the U.S. Department of the Interior, it maintains statistics on national and international energy supplies.

PICTURE CREDITS

PAGE

18: Grozaya/Shtterstock
39: Newscom
67: Newscom
97: Juan Fuertes/Shtterstock
121: Newscom

INDEX

INDEX

CONTRIBUTORS

PAUL RUSCHMANN, J.D., is a legal analyst and writer based in Canton, Michigan. He received his undergraduate degree from the University of Notre Dame and his law degree from the University of Michigan. He is a member of the State Bar of Michigan. His areas of specialization include legislation, public safety, traffic and transportation, and trade regulation. He is also the author of 12 other books in the POINT/COUNTERPOINT series, which deal with such issues as the military draft, indecency in the media, private property rights, the war on terror, and global warming. He can be found online at www.PaulRuschmann.com.

ALAN MARZILLI, M.A., J.D., lives in Birmingham, Alabama, and is a program associate with Advocates for Human Potential, Inc., a research and consulting firm based in Sudbury, Mass., and Albany, N.Y. He primarily works on developing training and educational materials for agencies of the federal government on topics such as housing, mental health policy, employment, and transportation. He has spoken on mental health issues in 30 states, the District of Columbia, and Puerto Rico; his work has included training mental health administrators, nonprofit management and staff, and people with mental illnesses and their families on a wide variety of topics, including effective advocacy, community-based mental health services, and housing. Marzilli has written several handbooks and training curricula that are used nationally and as far away as the U.S. territory of Guam. Additionally, he managed statewide and national mental health advocacy programs and worked for several public interest lobbying organizations while studying law at Georgetown University. Marzilli has written more than a dozen books, including numerous titles in the POINT/COUNTERPOINT series.